UNVEILING HAARP

UNVEILING HAARP

QUINN SILVER

CONTENTS

1 Chapter 1: Introduction to HAARP 1

2 Chapter 2: The Official Purpose of HAARP 7

3 Chapter 3: The Birth of Conspiracy Theories 15

4 Chapter 4: HAARP and Weather Control 23

5 Chapter 5: HAARP and Mind Control 31

6 Chapter 6: Media and Public Perception 39

7 Chapter 7: Scientific Community's Response 47

8 Chapter 8: Government and Military Involvement 55

9 Chapter 9: Case Studies of Alleged HAARP Interfere 65

10 Chapter 10: Debunking the Myths 75

11 Chapter 11: The Real Science Behind HAARP 85

12 Chapter 12: HAARP in Popular Culture 95

13 Chapter 13: Conclusion: Separating Fact from Ficti 105

14 Chapter 14: Appendices and References 111

Chapter 1: Introduction to HAARP

Origins and Objectives of HAARP

In the chilly expanse of Alaska, nestled near the small town of Gakona, lies a facility shrouded in both scientific ambition and public intrigue—the High-Frequency Active Auroral Research Program, or HAARP. This state-of-the-art installation, with its towering antennas and complex equipment, was born out of a vision to explore the unexplored realms of the ionosphere, the mysterious layer of the Earth's atmosphere that plays a crucial role in radio communication and navigation systems.

The journey of HAARP began in the early 1990s, driven by the collaborative efforts of the United States Air Force, the United States Navy, and the Defense Advanced Research Projects Agency (DARPA). The goal was to investigate the ionosphere's properties and behaviors, with the hope of harnessing this knowledge to enhance both civilian and military communication technologies. The project aimed to develop new methods for improving the reliability and range of radio signals, which are often disrupted by natural ionospheric conditions.

The inception of HAARP was not just a product of military interest; it also had significant academic backing. Researchers and scientists from various universities were intrigued by the potential of a facility that could systematically study the ionosphere. The University of Alaska Fairbanks became a key partner, contributing its expertise in geophysical sciences and providing a bridge between military objectives and academic research.

The primary objectives of HAARP were both broad and ambitious. They included understanding how the ionosphere impacts radio wave propagation, developing techniques for generating controlled plasma fields, and exploring the potential for creating artificial auroras. These goals were not only theoretically fascinating but also held practical implications for advancing communication systems, such as improving global positioning systems (GPS) and developing new technologies for over-the-horizon radar systems.

At its core, HAARP's mission was to unravel the complexities of the ionosphere, a region of the atmosphere that, despite its significance, remained largely enigmatic. By using high-frequency (HF) radio waves to stimulate small portions of the ionosphere, HAARP aimed to create controlled and predictable disturbances. These experiments provided scientists with valuable data, shedding light on the interactions between the ionosphere and radio waves.

As the project progressed, HAARP became a focal point for collaborative research, attracting scientists from around the world. It was a testament to human curiosity and the relentless pursuit of knowledge, driven by the desire to push the boundaries of what was known and to explore the possibilities that lay beyond the horizon.

Key Scientific Principles

Understanding the intricate workings of HAARP requires a basic grasp of the ionosphere and its critical role in our planet's communication infrastructure. The ionosphere is a region of the Earth's upper atmosphere, located approximately 30 to 600 miles above the surface. It is ionized by solar radiation, meaning it contains a significant number of free electrons and ions. This ionization affects radio wave propagation, making the ionosphere a crucial component for various forms of communication and navigation systems.

The ionosphere is divided into several layers, each with distinct characteristics. These layers—D, E, and F—vary in their electron density and their interaction with radio waves. The D layer is the lowest, absorbing high-frequency radio waves during the day and leading to signal attenuation. The E layer is found slightly higher and can refract radio waves back to Earth, enabling long-distance communication. The F layer, the highest of the three, is further divided into F1 and F2 regions. The F2 layer remains ionized even at night, supporting nighttime radio communication.

HAARP's scientific endeavors revolve around studying these layers, particularly the F region, which has a significant impact on long-distance radio communications. By using high-frequency (HF) transmitters, HAARP can stimulate small, controlled sections of the ionosphere. These transmitters send powerful radio waves into the ionosphere, creating temporary disturbances that researchers can study to understand how the ionosphere reacts and recovers.

One of the key scientific techniques employed by HAARP is ionospheric heating. This process involves directing intense HF radio waves at specific regions of the ionosphere, increasing the electron temperature and causing controlled perturbations. These perturbations can then be studied to gain insights into natural ionospheric processes and their effects on radio wave propagation. By observing the induced changes, scientists can develop models to predict how the ionosphere behaves under various conditions, ultimately leading to improved communication and navigation systems.

HAARP also utilizes a suite of instruments to measure the ionospheric responses to these controlled experiments. These instruments include ionosondes, which provide real-time data on the ionosphere's density and structure, and riometers, which measure cosmic noise absorption as an indicator of ionospheric conditions. Additionally, GPS receivers and magnetometers are used to monitor changes in ionospheric and geomagnetic activity.

The data collected from these experiments not only furthers our understanding of ionospheric physics but also has practical applications. For instance, by studying how the ionosphere affects GPS signals, researchers can develop methods to mitigate signal disruptions caused by ionospheric disturbances. This has significant implications for aviation, maritime navigation, and even everyday smartphone use.

In essence, HAARP's scientific principles and methodologies are designed to peel back the layers of the ionosphere, uncovering its mysteries and harnessing this knowledge to benefit both civilian and military communication technologies. Through a combination of advanced technology and methodical experimentation, HAARP continues to push the boundaries of what we know about this vital component of our planet's atmosphere.

History and Institutions Involved

The story of HAARP's development is one marked by technological ambition and interdisciplinary collaboration. From its conception to its current status as a prominent research facility,

HAARP has traversed a path filled with significant milestones, scientific achievements, and a fair share of public intrigue.

The foundation for HAARP was laid in 1990, spearheaded by the U.S. Air Force and Navy. These military branches envisioned a facility that could enhance understanding of the ionosphere to improve communication and surveillance technologies. The early 1990s saw the planning and construction phases, culminating in the groundbreaking of the HAARP Research Station in 1993 in Gakona, Alaska. This remote location was chosen for its quiet electromagnetic environment, ideal for conducting sensitive experiments without interference from urban radio signals.

The project's initial phase involved building the array of 180 high-frequency antennas, spread over a 33-acre field. These antennas, capable of transmitting powerful radio waves, formed the heart of HAARP's ability to stimulate and study the ionosphere. By 1999, the facility had reached its full operational capability, ready to conduct groundbreaking research.

Throughout its development, HAARP benefited from the involvement of several key institutions. The University of Alaska Fairbanks (UAF) played a crucial role, providing scientific expertise and acting as a bridge between military objectives and academic research. UAF's Geophysical Institute became a significant partner, contributing to the design and implementation of research projects. This partnership ensured that HAARP's work would not only serve military purposes but also advance scientific understanding of the ionosphere.

In addition to UAF, the project saw collaboration with several other universities and research organizations. Institutions like Stanford University and Penn State brought their own expertise to bear, contributing to HAARP's research agenda and helping to interpret the vast amounts of data generated by the facility. These academic partnerships underscored HAARP's dual mission of serving both national security interests and contributing to the broader scientific community.

HAARP's history is also marked by periods of public scrutiny and controversy. The facility's association with military research fueled speculation and conspiracy theories about its true purpose. Despite these challenges, HAARP continued to pursue its scientific goals, with transparency and public outreach becoming essential components of its operations.

Over the years, HAARP has achieved several notable scientific milestones. These include the successful generation of artificial auroras, advancements in understanding ionospheric dynamics, and contributions to the study of space weather phenomena. Each of these achievements has added to the body of knowledge about the ionosphere and its interaction with Earth's environment.

The facility has also faced challenges, including funding issues and operational changes. In 2014, ownership of HAARP was transferred to the University of Alaska Fairbanks, ensuring that the facility would continue to operate as a hub for ionospheric research. This transition marked a new chapter in HAARP's history, emphasizing its role in academic research and public scientific inquiry.

Today, HAARP stands as a testament to the power of interdisciplinary collaboration and the pursuit of knowledge. Its history is a complex tapestry of military ambition, scientific curiosity, and public interest, reflecting the multifaceted nature of modern research initiatives. As HAARP continues to explore the frontiers of ionospheric science, it remains a symbol of the enduring quest to understand the natural world and harness its mysteries for the benefit of humanity.

Chapter 2: The Official Purpose of HAARP

Ionospheric Research

The High-Frequency Active Auroral Research Program, or HAARP, stands as a monument to scientific inquiry and technological advancement. At its core, HAARP's primary mission is to explore and understand the ionosphere—a layer of Earth's atmosphere that is crucial for radio communication, navigation, and many other technological applications.

The ionosphere is a region of the upper atmosphere that extends from about 30 miles to 600 miles above the Earth. It is ionized by solar radiation, which means it contains a high concentration of free electrons and ions. This ionization affects how radio waves travel through the atmosphere, making the ionosphere an essential component in the global communication network. Understanding the ionosphere's properties and behavior is vital for improving communication technologies and mitigating the impacts of space weather on these systems.

One of the primary scientific objectives of HAARP is to study how the ionosphere responds to both natural and artificial stimuli. This research helps scientists develop models to predict ionospheric conditions and their effects on radio wave propagation. By doing so, HAARP contributes to a better understanding of how to improve the reliability and effectiveness of communication systems that rely on the ionosphere.

To achieve these objectives, HAARP employs high-frequency (HF) transmitters to stimulate small, controlled regions of the ionosphere. These transmitters generate powerful radio waves that are directed into the ionosphere, creating temporary disturbances. Scientists then study these disturbances to understand how the ionosphere reacts and recovers. This process, known as ionospheric heating, allows researchers to simulate natural phenomena, such as solar flares and geomagnetic storms, under controlled conditions.

HAARP's research methods are as sophisticated as they are varied. The facility utilizes a range of instruments to measure the ionosphere's responses to these controlled experiments. Ionosondes, for example, provide real-time data on the ionosphere's density and structure by transmitting radio waves and measuring their reflection. Riometers measure cosmic noise absorption, which indicates changes in ionospheric conditions. GPS receivers and magnetometers monitor variations in ionospheric and geomagnetic activity, providing comprehensive data for analysis.

The insights gained from HAARP's ionospheric research have far-reaching implications. For instance, understanding how the ionosphere affects GPS signals can lead to methods for mitigating signal disruptions, which is crucial for aviation, maritime navigation, and even everyday smartphone use. By improving the accuracy and reliability of these systems, HAARP's research benefits both civilian and military applications.

Furthermore, HAARP's studies on ionospheric behavior contribute to our knowledge of space weather—a term used to describe the changing environmental conditions in space that can affect Earth and its technological systems. Space weather events, such as solar flares and geomagnetic storms, can disrupt communication and navigation systems, power grids, and satellite operations. By simulating these conditions and studying their effects, HAARP helps develop strategies to protect these critical systems from space weather impacts.

In summary, HAARP's ionospheric research is a cornerstone of its mission. By unraveling the mysteries of the ionosphere and its interaction with radio waves, HAARP not only advances our scientific knowledge but also enhances the technologies that underpin modern communication and navigation systems. This pursuit of understanding and innovation underscores HAARP's significance in both the scientific community and the broader technological landscape.

Advancements in Radio Communications

The High-Frequency Active Auroral Research Program (HAARP) isn't just about unraveling the mysteries of the ionosphere—it's also at the forefront of advancing radio communication technologies. The ionosphere plays a critical role in the transmission of radio waves, and understanding its complex behavior is essential for improving the reliability and effectiveness of radio communication systems worldwide.

One of the main challenges in radio communications is the variability of the ionosphere. Natural phenomena such as solar flares, geomagnetic storms, and even daily changes in solar radiation can cause significant disruptions to radio signals. These disruptions can degrade the quality of radio communications, leading to static, signal loss, and even complete communication blackouts. For industries and services that rely heavily on radio communications, such as aviation, maritime operations, and emergency response, these disruptions can have serious implications.

HAARP's research into the ionosphere is designed to address these challenges. By understanding how the ionosphere affects radio wave propagation, HAARP aims to develop new techniques to improve signal quality and reliability. One key aspect of this research involves studying the interaction between high-frequency (HF) radio waves and the ionosphere. HAARP's powerful HF transmitters can create controlled disturbances in the ionosphere, allowing scientists to observe and analyze the resulting changes in radio wave propagation.

These experiments have led to several advancements in radio communication technology. For example, HAARP's research has contributed to the development of over-the-horizon radar systems. These radar systems use ionospheric reflection to detect objects beyond the line of sight, making them invaluable for military surveillance and tracking. By understanding how the ionosphere affects radar signals, HAARP has helped improve the accuracy and range of these systems.

Another significant advancement driven by HAARP's research is the enhancement of global positioning systems (GPS). GPS signals, which travel through the ionosphere, are susceptible to ionospheric disturbances that can cause inaccuracies in positioning data. By studying these disturbances, HAARP has provided valuable insights into how to mitigate their effects, leading to more accurate and reliable GPS systems. This is particularly important for aviation and maritime navigation, where precise positioning is crucial for safety and efficiency.

HAARP's contributions to radio communication technologies extend beyond military applications. The research conducted at HAARP has practical benefits for civilian communication systems as well. For instance, improved understanding of ionospheric conditions can help enhance the reliability of radio broadcasting, amateur radio operations, and emergency communication networks. By developing techniques to compensate for ionospheric variability, HAARP helps ensure that these systems remain robust and dependable, even in the face of space weather events.

In addition to its technological advancements, HAARP's work in radio communications underscores the importance of interdisciplinary collaboration. The facility brings together scientists and engineers from various fields, including atmospheric science, physics, and electrical engineering. This collaborative approach allows HAARP to tackle complex problems from multiple angles, leading to innovative solutions that benefit a wide range of communication technologies.

In conclusion, HAARP's research into the ionosphere and its impact on radio wave propagation has led to significant advancements in radio communication technologies. By addressing the challenges posed by ionospheric variability, HAARP helps improve the reliability and effectiveness of communication systems that are vital for both civilian and military applications. This work not only enhances our understanding of the ionosphere but also underscores the importance of continued scientific exploration and technological innovation.

Scientific Exploration and Technological Methodology

The High-Frequency Active Auroral Research Program (HAARP) is a testament to the synergy of cutting-edge technology and scientific ambition. Beyond its primary objectives of ionospheric research and advancements in radio communications, HAARP embodies the spirit of scientific exploration and the relentless pursuit of knowledge. This section delves into HAARP's broader scientific exploration goals and the sophisticated technological methodologies that underpin its operations.

One of HAARP's distinguishing features is its interdisciplinary approach to scientific exploration. The facility serves as a nexus where physics, engineering, and atmospheric science converge. This interdisciplinary collaboration allows HAARP to tackle complex scientific questions from multiple angles, leading to comprehensive and innovative solutions. For instance, HAARP's research into the ionosphere not only advances atmospheric science but also contributes to our understanding of space weather phenomena and their impact on Earth's environment.

A key aspect of HAARP's scientific exploration is its ability to generate controlled ionospheric perturbations. By using high-frequency (HF) radio waves to stimulate specific regions of the ionosphere, HAARP creates temporary disturbances that can be studied in detail. These controlled experiments allow scientists to observe and measure the ionosphere's responses to various stimuli, providing valuable insights into its behavior and properties. This ability to simulate natural phenomena under controlled conditions is one of HAARP's greatest strengths, enabling researchers to test hypotheses and develop predictive models.

The technological methodology employed by HAARP is equally impressive. At the heart of the facility is the HF antenna array, a sprawling field of 180 antennas capable of transmitting powerful radio waves into the ionosphere. These antennas are arranged in a grid pattern, allowing for precise control over the direction and intensity of the transmitted signals. This level of control is crucial for conducting targeted experiments and ensuring the accuracy of the resulting data.

In addition to the HF transmitters, HAARP utilizes a suite of advanced instruments to monitor and measure ionospheric conditions. Ionosondes, for example, transmit radio pulses and measure their reflection from the ionosphere, providing real-time data on its density and structure. Magnetometers measure changes in the Earth's magnetic field, offering insights into geomagnetic activity and its interaction with the ionosphere. Optical instruments, such as all-sky cameras, capture images of auroras and other visual phenomena, helping scientists correlate visual observations with ionospheric data.

The data collected from these instruments is meticulously analyzed to extract meaningful insights. Researchers use sophisticated software and computational models to process the data and identify patterns and trends. This analytical process is integral to HAARP's mission, as it allows scientists to refine their understanding of the ionosphere and develop accurate predictive models. These models, in turn, have practical applications in improving communication and navigation systems.

Looking to the future, HAARP's potential for scientific exploration remains vast. Ongoing and future research projects aim to push the boundaries of what is known about the ionosphere and its interaction with Earth's environment. For example, scientists are exploring the potential for using HAARP to study space weather phenomena, such as solar flares and their impact on the ionosphere. This research could lead to new strategies for protecting critical infrastructure from space weather-related disruptions.

In conclusion, HAARP represents a remarkable fusion of scientific exploration and technological innovation. Through its interdisciplinary approach, advanced methodologies, and commitment to understanding the ionosphere, HAARP continues to make significant contributions to both fundamental science and practical applications. As a beacon of scientific inquiry, HAARP embodies the relentless pursuit of knowledge and the quest to unlock the mysteries of our planet's atmosphere.

Chapter 3: The Birth of Conspiracy Theories

The Origins of HAARP-Related Conspiracy Theories

The High-Frequency Active Auroral Research Program (HAARP) has long been a source of intrigue and suspicion, giving rise to numerous conspiracy theories since its inception. Understanding the origins of these theories requires a look at the socio-political context of the time and the factors that fueled early suspicions.

In the early 1990s, HAARP was established amidst a backdrop of growing public mistrust toward governmental and military projects. This was a period marked by the end of the Cold War, during which secrecy and high-stakes technological advancements were the norm. The announcement of a new, advanced research facility in the remote wilderness of Alaska naturally piqued public curiosity and, with it, suspicion.

The facility's mission—to study the ionosphere using high-frequency radio waves—was shrouded in scientific jargon, making it difficult for the general public to grasp its true purpose. This lack of transparency created a fertile ground for speculation. Early suspicions were often rooted in a general mistrust of the government and military, coupled with fears about new and unfamiliar technologies.

The public's imagination ran wild with possibilities, and soon, the notion that HAARP was more than just a research facility began to take hold.

Misinformation played a significant role in the birth of HAARP-related conspiracy theories. In an era before the widespread use of the internet, rumors and speculative ideas spread through word of mouth, amateur publications, and fringe media. Early rumors suggested that HAARP's powerful radio transmitters could manipulate the weather, control minds, or even cause natural disasters. These ideas were often fueled by a lack of credible information and the mysterious nature of the project's operations.

One of the earliest speculative ideas was that HAARP could be used for weather control. This theory proposed that the facility's radio waves could manipulate the atmosphere, creating artificial weather patterns. The notion of weather control was not entirely new—historical attempts at cloud seeding and other forms of weather modification had been documented—but the scale and secrecy of HAARP added a new layer of intrigue. As these rumors spread, they began to merge with other fringe theories about government control and technological manipulation.

Another pervasive idea was that HAARP could be used for mind control. This theory suggested that the facility's powerful radio waves could influence human thought and behavior. The concept of mind control, while far-fetched, tapped into deep-seated fears about government overreach and the potential misuse of technology. Early proponents of this theory often pointed to HAARP's military connections and its location in the remote Alaskan wilderness as evidence of a nefarious agenda.

The birth of HAARP-related conspiracy theories can also be attributed to the broader cultural and historical context of the time. The early 1990s saw a rise in conspiracy culture, driven by a growing distrust of authority and an increasing appetite for alternative explanations. This cultural shift was reflected in popular media, with television shows, movies, and books exploring themes of government secrecy, alien encounters, and technological experimentation. In this environment, HAARP became a convenient focal point for a multitude of fears and suspicions.

In summary, the origins of HAARP-related conspiracy theories lie in a combination of socio-political factors, misinformation, and cultural trends. The project's secrecy, coupled with a lack of public understanding and a growing appetite for conspiracy theories, created the perfect storm for the birth of these ideas. As we explore the key figures and moments that contributed to the development of these theories, we will gain a deeper understanding of how HAARP became a lightning rod for public suspicion and intrigue.

Key Figures and Moments in the Development of Conspiracy Theories

The evolution of HAARP-related conspiracy theories has been shaped by a cast of influential figures, pivotal moments, and the power of media in amplifying suspicions and spreading misinformation. To understand how these theories gained traction, it is essential to examine the key personalities who propagated these ideas and the significant events that fueled their dissemination.

One of the most prominent figures in the propagation of HAARP conspiracy theories is Nick Begich, an author and lecturer who co-authored the book *"Angels Don't Play This HAARP."* Published in the mid-1990s, this book became a seminal work for those who believed in the darker purposes of HAARP. Begich, along with his co-author Jeane Manning, argued that HAARP could be used for purposes far beyond its stated scientific goals, including weather modification and mind control. Their claims, though controversial and largely dismissed by the scientific community, resonated with a segment of the public that was already predisposed to mistrust government and military projects.

The media played a crucial role in amplifying these theories. Television documentaries and radio shows dedicated to uncovering "hidden truths" frequently featured segments on HAARP, often presenting speculative information as fact. For instance, the History Channel's popular show *"Conspiracy Theories with Jesse Ventura"* aired an episode that delved into HAARP's alleged capabilities. Hosted by former professional wrestler and governor Jesse Ventura, the show gave a platform to theorists who claimed that HAARP could manipulate weather patterns, control minds, and even induce earthquakes.

Online platforms and forums also became breeding grounds for HAARP conspiracy theories. Websites dedicated to alternative news and conspiracy theories provided a space for like-minded individuals to share their beliefs and "evidence." Social media further amplified these ideas, allowing them to reach a global audience and gain a veneer of credibility through repetition and viral spread. Figures like Alex Jones, a prominent conspiracy theorist, used his platform Infowars to discuss HAARP's supposed nefarious activities, influencing millions of listeners and viewers.

Pivotal moments in history also played a significant role in the rise of HAARP-related conspiracy theories. Natural disasters and unusual weather events were often linked to HAARP by conspiracy theorists looking for explanations that fit their narrative. For example, the devastating 2010 earthquake in Haiti was quickly attributed to HAARP by some theorists, who claimed that the facility's experiments had triggered the quake. Similarly, unusual weather patterns, such as unseasonal storms or extreme temperatures, were often cited as evidence of HAARP's weather manipulation capabilities.

These conspiracy theories persisted and evolved, adapting to new information and technological advancements. Each time HAARP conducted a new experiment or when new scientific findings were published, theorists interpreted these developments through a lens of suspicion and mistrust. Despite numerous debunking efforts by scientists and researchers, the core beliefs about HAARP's malevolent purposes remained steadfast in the minds of many.

The psychological and social factors that made these theories appealing are multifaceted. Conspiracy theories often provide simple explanations for complex phenomena, offering a sense of control and understanding in an uncertain world. They also reinforce preexisting beliefs and biases, creating a feedback loop that is difficult to break. For those who distrusted government and military institutions, HAARP became a symbol of hidden agendas and technological overreach.

In summary, the development of HAARP-related conspiracy theories was shaped by influential figures, media amplification, and significant historical events. These factors combined to create a potent mix of suspicion, misinformation, and public intrigue that has endured despite scientific efforts to dispel these myths. As we examine the evolution and impact of these theories, we will gain further insight into their persistent appeal and the challenges they pose to scientific communication and public trust.

The Evolution and Impact of HAARP Conspiracy Theories

The evolution of HAARP-related conspiracy theories has been a dynamic process, adapting and morphing in response to new information, technological advancements, and shifting public senti-ments. Despite numerous debunking efforts by scientists and researchers, these theories have persisted and evolved, continuously capturing the public's imagination and influencing perceptions of HAARP and similar scientific endeavors.

One of the key factors in the evolution of these theories is their ability to adapt to new scientific information and technological ad-vancements. Each time HAARP conducted a new experiment or published new findings, conspiracy theorists interpreted these devel-opments through a lens of suspicion. For instance, when HAARP released data on artificial auroras generated through its experiments, theorists claimed it was proof of weather manipulation or mind con-trol. This ability to reinterpret scientific data to fit pre-existing be-liefs has allowed HAARP conspiracy theories to remain relevant and compelling to their adherents.

The persistence of certain core beliefs, despite efforts to debunk them, is another hallmark of the evolution of these theories. Central ideas, such as HAARP's alleged capabilities in weather control and mind control, have remained steadfast. Even when confronted with scientific evidence that disproves these claims, theorists often dismiss such evidence as part of a cover-up or misinformation campaign. This unwavering belief in the core tenets of HAARP conspiracy the-ories underscores the psychological appeal of these narratives—they offer simple explanations for complex phenomena and reinforce dis-trust in authoritative institutions.

The impact of HAARP conspiracy theories on public percep-tion has been profound. These theories have significantly influenced how people view not only HAARP but also other scientific and governmental projects. For many, HAARP symbolizes the hidden dangers of advanced technology and the potential for misuse by powerful entities. This perception has been exacerbated by the proliferation of these theories through social media and online platforms, where misinformation can spread rapidly and gain credibility through repetition.

Social media, in particular, has played a pivotal role in perpetuating and evolving HAARP conspiracy theories. Platforms like YouTube, Facebook, and Twitter have become hotbeds for the dissemination of alternative narratives and fringe theories. The algorithms that drive these platforms often prioritize engaging and sensational content, which can lead to the amplification of conspiracy theories. As a result, HAARP-related content frequently appears in recommendations and feeds, reaching a wide audience and perpetuating the cycle of misinformation.

The consequences of HAARP conspiracy theories extend beyond mere public perception. These theories have real-world implications for scientific research and policy. Public fear and mistrust, fueled by conspiracy theories, can hinder scientific progress and complicate the implementation of important projects. For instance, opposition to scientific initiatives, based on unfounded fears, can lead to delays, increased costs, and even the cancellation of beneficial research. This highlights the broader challenge of combating misinformation and promoting scientific literacy.

Efforts to debunk HAARP conspiracy theories and restore public trust have been ongoing. Scientists and researchers have engaged in public outreach, providing transparent information about HAARP's objectives and methodologies. Educational initiatives aim to improve public understanding of ionospheric research and the scientific process. Despite these efforts, the entrenched nature of conspiracy beliefs poses a significant challenge. Building public trust requires not only debunking myths but also fostering a culture of critical thinking and scientific inquiry.

In conclusion, the evolution and impact of HAARP conspiracy theories illustrate the complex interplay between scientific advancement, public perception, and misinformation. These theories have adapted to new information and technological changes, maintaining their appeal and influence over time. Addressing the challenges posed by these theories requires a multifaceted approach, involving transparent communication, public education, and a commitment to scientific integrity. As HAARP continues its research, the dialogue between scientists and the public remains crucial in bridging the gap between fact and fiction.

Chapter 4: HAARP and Weather Control

Claims of Weather Control

The theory that HAARP is capable of controlling the weather is perhaps the most enduring and widespread of all the conspiracy theories associated with the facility. This theory has captivated the imagination of many, fueled by a combination of scientific misunderstanding, mistrust in government, and sensationalist media coverage. To understand the claims of weather control attributed to HAARP, we must explore the origins of this theory, the specific mechanisms proposed by conspiracy theorists, and notable events that have been linked to HAARP's supposed capabilities.

The origins of the weather control theory can be traced back to the early days of HAARP's operation in the 1990s. During this period, global awareness of climate change and unusual weather patterns was growing, and there was a burgeoning interest in the possibility of weather modification. HAARP's remote location in Alaska, combined with its association with the U.S. military, provided a perfect breeding ground for suspicion and speculation. The idea that a hidden facility could wield the power to alter weather patterns resonated with the public's preexisting fears of government overreach and technological manipulation.

Conspiracy theorists argue that HAARP uses its high-frequency radio waves to manipulate the ionosphere in ways that can influence weather patterns. One of the most frequently cited methods is the creation of artificial clouds or storms. Theorists claim that by heating specific areas of the ionosphere, HAARP can cause changes in atmospheric pressure, leading to the formation of clouds and precipitation. Some even suggest that HAARP can generate hurricanes or steer weather systems to target specific regions, citing unusual weather events as evidence of these capabilities.

Another aspect of the weather control theory involves the manipulation of jet streams. Jet streams are fast-flowing air currents in the upper levels of the atmosphere that have a significant impact on weather patterns. Conspiracy theorists believe that HAARP's radio waves can alter the flow of these jet streams, redirecting weather systems to create droughts, floods, or extreme temperatures. This idea is often supported by pointing to anomalous weather patterns and extreme weather events that seem to deviate from typical seasonal variations.

Notable weather events have frequently been attributed to HAARP by those who subscribe to this theory. For example, Hurricane Katrina, which devastated the Gulf Coast of the United States in 2005, has been cited as a potential HAARP-induced event. Theorists argue that the storm's intensity and the unusual atmospheric conditions surrounding it were indicative of weather manipulation. Similarly, the 2010 heatwave in Russia, which caused widespread drought and wildfires, has been linked to HAARP's supposed ability to control weather, with claims that the facility was used to create or exacerbate the extreme conditions.

The idea that HAARP could be used for weather control taps into a deeper, primal fear of humanity's vulnerability to natural forces. The unpredictability and often destructive power of the weather make it a potent subject for conspiracy theories, especially when combined with the notion of secretive government projects. For many, the idea that a hidden hand could be directing these natural phenomena is both terrifying and oddly comforting, providing a clear target for their fears and frustrations.

In summary, the claims of weather control attributed to HAARP are rooted in a mix of scientific misunderstanding, distrust of government, and the inherent drama of weather-related disasters. These theories have been fueled by notable weather events and have evolved to include increasingly sophisticated mechanisms of manipulation. As we delve into the scientific counterarguments to these claims, we will see how the reality of HAARP's capabilities stands in stark contrast to the sensationalist narratives that surround it.

Scientific Counterarguments

While the idea that HAARP can control the weather is captivating, it is essential to address these claims through the lens of scientific scrutiny. Theories suggesting that HAARP can manipulate weather patterns lack credible scientific support and are often based on a misunderstanding of HAARP's actual capabilities and the scientific principles governing weather and the ionosphere.

To begin with, it is important to understand the actual scientific capabilities of HAARP. The facility is equipped with high-frequency (HF) transmitters designed to study the ionosphere, a layer of Earth's atmosphere ionized by solar radiation. HAARP's primary objective is to investigate how the ionosphere affects radio wave propagation, with the goal of improving communication and navigation systems. The energy output of HAARP's transmitters, while significant, is minuscule compared to the energy required to influence weather patterns on a global or even regional scale.

Weather modification involves complex interactions between the atmosphere, hydrosphere, and various climatic factors. Techniques such as cloud seeding, which aims to enhance precipitation by dispersing substances into the air that serve as cloud condensation or ice nuclei, have been studied and implemented to a limited extent. However, these techniques require specific atmospheric conditions and cannot produce large-scale or long-lasting weather changes. The notion that HAARP's HF transmitters could generate enough energy to create or control weather events like hurricanes, droughts, or significant storms is not supported by scientific evidence.

Moreover, weather phenomena are driven by massive energy exchanges in the atmosphere. For example, hurricanes derive their energy from the warm waters of the ocean, and the amount of energy involved is far beyond the capability of any human-made technology. HAARP's transmitters, designed to perturb small, localized areas of the ionosphere for short periods, simply do not possess the capacity to influence these immense natural processes.

Scientific experts have consistently refuted the claims of HAARP's weather control capabilities. Numerous studies and expert opinions have emphasized that HAARP's activities are focused on ionospheric research and do not have the power to affect weather systems. For instance, Dr. John Heckscher, a HAARP program manager at the Air Force Research Laboratory, has stated that the facility's operations are limited to scientific exploration of the ionosphere and cannot be used for weather modification. Similarly, experts from the Geophysical Institute at the University of Alaska Fairbanks, which operates HAARP, have clarified that the energy levels used in HAARP experiments are insufficient to impact weather patterns.

Debunking specific claims made by theorists is also crucial. For example, the idea that HAARP can create hurricanes is scientifically unfounded. Hurricanes are complex systems that develop over warm ocean waters and are driven by vast amounts of heat and moisture. The energy required to generate a hurricane is on the order of hundreds of terawatts, which is many magnitudes greater than HAARP's output. Similarly, the notion that HAARP could cause earthquakes by altering tectonic stress through ionospheric manipulation is not supported by geological science. Earthquakes result from the movement of tectonic plates and the release of stress accumulated over time, processes that are entirely unrelated to ionospheric research.

In conclusion, the scientific counterarguments to the claims of HAARP's weather control capabilities are robust and well-founded. HAARP's purpose and capabilities are centered on ionospheric research, and its energy output is insufficient to influence weather systems. Understanding the scientific principles behind weather modification and HAARP's actual operations helps dispel the myths and underscores the importance of evidence-based science in debunking conspiracy theories.

Notable Weather Events Attributed to HAARP

The High-Frequency Active Auroral Research Program (HAARP) has been implicated in numerous weather-related conspiracy theories, with some theorists attributing specific natural disasters and unusual weather patterns to HAARP's alleged capabilities. This section examines notable weather events that have been linked to HAARP, analyzes the theories behind these attributions, and provides scientific explanations for these phenomena.

One of the most frequently cited examples by conspiracy theorists is Hurricane Katrina, which struck the Gulf Coast of the United States in August 2005. The hurricane caused catastrophic damage and loss of life, and its aftermath led to widespread speculation about its origins. Some theorists claimed that HAARP was used to intensify the storm and direct it toward New Orleans, arguing that the unusual strength and rapid intensification of Katrina were evidence of artificial manipulation. However, scientific analysis shows that Katrina's development and path were consistent with known meteorological patterns. Hurricanes form over warm ocean waters, and Katrina benefited from exceptionally warm Gulf waters and favorable atmospheric conditions that fueled its intensity. There is no scientific basis for the claim that HAARP influenced the hurricane.

Another event often linked to HAARP is the 2010 Russian heatwave. That summer, Russia experienced unprecedented temperatures, leading to severe drought, wildfires, and significant loss of life. Conspiracy theorists suggested that HAARP was behind this extreme weather, pointing to the unusual nature of the heatwave as evidence. However, climate scientists attribute the heatwave to a combination of natural atmospheric phenomena, including a strong blocking high-pressure system that prevented cooler air from moving into the region. Such heatwaves, while extreme, can occur due to natural variations in weather patterns and are not indicative of artificial interference.

The 2011 Tōhoku earthquake and tsunami in Japan, which caused devastating damage and triggered the Fukushima nuclear disaster, has also been cited by theorists as a HAARP-induced event. Some claim that the ionospheric disturbances observed before the earthquake were a result of HAARP's activities. However, geophysicists have thoroughly studied the earthquake and found that it was caused by the release of stress along a subduction zone where the Pacific Plate is being forced under the North American Plate. The ionospheric disturbances observed were actually a result of the massive energy release from the earthquake itself, not a precursor caused by HAARP. Earthquakes are driven by tectonic processes deep within the Earth's crust and are not influenced by atmospheric or ionospheric conditions.

These examples highlight the tendency of conspiracy theories to link HAARP to notable natural disasters and unusual weather events. This tendency often stems from a misunderstanding of the natural processes involved in these phenomena and a mistrust of scientific explanations. The inclination to seek alternative explanations for extreme events can be influenced by the fear and uncertainty these events generate. However, attributing them to HAARP lacks credible scientific evidence and diverts attention from the real causes and solutions.

Scientific explanations for these events emphasize the importance of understanding natural atmospheric, oceanic, and geological processes. Weather patterns are driven by complex interactions between the Earth's atmosphere, oceans, and landmasses, while earthquakes result from the movement of tectonic plates. These processes are studied extensively by scientists, who use data and models to predict and explain such phenomena. Engaging with this scientific knowledge is crucial for developing effective responses to natural disasters and mitigating their impacts.

In conclusion, the attribution of notable weather events and natural disasters to HAARP by conspiracy theorists is not supported by scientific evidence. Understanding the real causes of these phenomena requires a reliance on established scientific principles and rigorous research. As we continue to explore HAARP's actual capabilities and limitations, it becomes clear that these theories are based on a combination of misinformation, fear, and misunderstanding.

Chapter 5: HAARP and Mind Control

The Basis for Mind Control Claims

The theory that HAARP is capable of mind control is one of the most sensational and controversial aspects of the conspiracy narratives surrounding the facility. To understand the origins and foundations of these claims, it is essential to explore the historical context, the sources of these assertions, and the initial proponents who brought this theory to public attention.

The socio-political environment in which HAARP was established played a significant role in the development of mind control theories. The early 1990s were a time of significant global change, marked by the end of the Cold War and a shift in the geopolitical landscape. This period was also characterized by a growing public awareness and fear of government surveillance and control, fueled by revelations of past covert operations and experiments. Notable among these were the CIA's MKUltra program, which involved unethical experiments on mind control and behavior modification. These historical events created a fertile ground for suspicion and fear, making the idea of government-run mind control projects like HAARP more plausible to the public.

The theory that HAARP could be used for mind control gained traction through the efforts of a few key individuals. One of the most prominent early proponents was Nick Begich, who co-authored the book "Angels Don't Play This HAARP." In this book, Begich and his co-author Jeane Manning argued that HAARP's high-frequency radio waves could be used to manipulate human thoughts and emotions. Begich's background in politics and his outspoken nature helped bring these ideas to a wider audience. His book and subsequent lectures introduced the notion of HAARP as a tool for psychological manipulation, blending scientific jargon with speculative theories.

Another influential figure in the propagation of mind control theories was Dr. Rosalie Bertell, an environmental scientist and activist. Bertell expressed concerns about the potential military applications of HAARP and its possible effects on human health and behavior. Her status as a respected scientist lent credibility to the claims, despite the lack of empirical evidence supporting the mind control assertions. Bertell's involvement highlighted how legitimate concerns about environmental and public health impacts could be co-opted into more sensational conspiracy theories.

The alleged methods of mind control attributed to HAARP revolve around the use of electromagnetic waves and radio frequencies. Conspiracy theorists claim that HAARP's powerful transmissions can be used to influence brainwaves, alter mental states, and control behavior. They argue that by targeting specific frequencies, HAARP can induce feelings of anxiety, depression, or even manipulate thoughts directly. These theories often cite the work of scientists like Nikola Tesla, whose experiments with wireless energy transmission are reinterpreted as precursors to modern mind control technologies.

These claims are supported by a misunderstanding of scientific principles and a misinterpretation of legitimate research. For example, theorists often reference the phenomenon of electromagnetic radiation affecting biological tissues, which is a well-documented field of study. However, the leap from understanding these interactions to asserting that HAARP can control minds is not supported by scientific evidence. The complexity of human cognition and behavior, which involves intricate neural networks and biochemical processes, cannot be manipulated by the relatively low-energy radio waves produced by HAARP.

In conclusion, the basis for the mind control claims associated with HAARP is rooted in historical fears, influential proponents, and a misunderstanding of scientific principles. The socio-political context of the 1990s, combined with past instances of unethical government experiments, created an environment ripe for the emergence of such theories. As we examine the alleged methods of mind control and the scientific rebuttals to these claims, it becomes clear that the reality of HAARP's capabilities is far removed from the sensational narratives that have captured the public's imagination.

Alleged Methods of Mind Control

The theory that HAARP is used for mind control hinges on several alleged methods that conspiracy theorists claim the facility employs to influence human cognition and behavior. These methods are often presented with a veneer of scientific terminology, borrowing concepts from legitimate research but distorting them to fit the narrative of HAARP as a tool for psychological manipulation. To understand these claims, it is important to examine the specific technologies and mechanisms purportedly involved, as well as the scientific principles behind them.

Conspiracy theorists argue that HAARP's powerful high-frequency (HF) transmitters can emit electromagnetic waves and radio frequencies capable of influencing the human brain. According to these theories, HAARP can target specific frequencies that correspond to brainwave patterns, thereby altering thoughts, emotions, and behaviors. The primary technologies implicated in these claims include electromagnetic radiation, radio frequency modulation, and brainwave entrainment.

Electromagnetic radiation is a broad term that encompasses various types of waves, including radio waves, microwaves, and X-rays. It is well-established that certain types of electromagnetic radiation can affect biological tissues. For example, exposure to high levels of microwave radiation can cause heating of tissues, while ultraviolet radiation from the sun can damage skin cells. However, the specific claim that HAARP's HF transmitters can emit radiation capable of directly influencing brain function lacks scientific support. The energy levels and frequencies used by HAARP are insufficient to penetrate the skull and affect brain activity in any significant way.

Radio frequency modulation is another technique mentioned by theorists. This involves varying the frequency or amplitude of radio waves to encode information. In the context of HAARP, theorists claim that modulating radio frequencies can influence brainwave patterns. Brainwaves are electrical impulses in the brain that occur at different frequencies, typically measured in Hertz (Hz). Common brainwave frequencies include alpha waves (8-12 Hz), beta waves (13-30 Hz), and theta waves (4-7 Hz). While it is true that brainwave patterns can be influenced by external stimuli, such as light and sound, the idea that radio frequency modulation can exert precise control over brainwaves is not supported by scientific evidence. The brain's electromagnetic environment is complex, and the external signals would need to be extremely precise and powerful to have any meaningful impact.

Brainwave entrainment is a concept often cited in support of mind control theories. This phenomenon occurs when external stimuli, such as rhythmic light or sound, synchronize with the brain's natural frequencies, potentially influencing mental states. For example, listening to binaural beats—sounds played at slightly different frequencies in each ear—can induce relaxation or concentration. However, the claim that HAARP can use this principle to control minds is speculative at best. The facility's HF transmissions are not designed to produce the specific frequencies or patterns necessary for brainwave entrainment, and even if they were, the complexity of human cognition would make precise control highly unlikely.

In summary, the alleged methods of mind control attributed to HAARP involve a combination of misunderstood scientific concepts and speculative theories. While electromagnetic radiation, radio frequency modulation, and brainwave entrainment are legitimate areas of research, their application to mind control through HAARP is not supported by scientific evidence. The energy levels, frequencies, and mechanisms involved in HAARP's operations are not capable of influencing human cognition and behavior in the ways suggested by conspiracy theorists. As we turn to scientific rebuttals and evidence, we will further debunk these claims and highlight the importance of critical thinking and scientific literacy in evaluating such theories.

Scientific Rebuttals and Evidence

The claims that HAARP is involved in mind control through the use of electromagnetic waves and radio frequencies have been consistently debunked by the scientific community. These assertions, often rooted in a misunderstanding of both HAARP's purpose and the science of electromagnetic effects on the human body, do not hold up under scientific scrutiny. In this section, we will examine expert opinions, peer-reviewed studies, and the broader scientific consensus to rebut these mind control theories.

Firstly, it is important to highlight the statements and findings from credible scientists who have studied HAARP and its effects. Dr. John Heckscher, a former program manager at HAARP and a senior research physicist, has explicitly stated that the facility is incapable of affecting human health or behavior. According to Heckscher, the energy levels produced by HAARP are far too low to have any impact on the brain or nervous system. HAARP's primary objective is to study the ionosphere for advancements in radio communications and navigation, not to manipulate human thought processes.

Similarly, researchers at the Geophysical Institute at the University of Alaska Fairbanks, which operates HAARP, have echoed these sentiments. They emphasize that HAARP's experiments are focused on understanding natural atmospheric phenomena and improving communication technologies. The scientific community at large does not support the idea that HAARP could be used for mind control, as the facility's design and operational parameters do not align with the capabilities required for such activities.

Peer-reviewed studies on the effects of electromagnetic fields (EMFs) and radio frequencies (RFs) on human health further dismantle the mind control theory. Extensive research has been conducted to evaluate the potential health impacts of EMFs and RFs, particularly from sources like cell phones, microwave ovens, and power lines. While high levels of certain types of radiation can be harmful, the frequencies and power levels used by HAARP are not in the range that would affect human health.

For instance, the World Health Organization (WHO) and numerous other health agencies have conducted comprehensive reviews of the scientific literature on EMF exposure. These reviews consistently find no conclusive evidence that low-level EMF exposure, such as that from HAARP, poses a risk to human health. The energy levels involved in HAARP's ionosphcric experiments are many orders of magnitude lower than those required to influence neural activity or behavior.

Addressing specific claims made by conspiracy theorists is also crucial. The assertion that HAARP can influence thoughts or emotions through electromagnetic waves is not supported by any credible scientific evidence. The human brain operates through complex electrochemical processes that are not easily influenced by external electromagnetic fields at the levels HAARP operates. Moreover, the brain's natural electromagnetic environment is highly regulated and robust against low-level external perturbations.

The role of scientific literacy in evaluating and understanding these claims cannot be overstated. A basic understanding of physics, biology, and the scientific method is essential for critically assessing the validity of such theories. Educators and scientists have a vital role in improving public understanding of these concepts, helping to dispel myths and combat misinformation. Critical thinking and skepticism are necessary tools in distinguishing between scientifically grounded information and pseudoscientific theories.

In conclusion, the scientific rebuttals and evidence against the mind control claims related to HAARP are clear and compelling. HAARP's capabilities and operations are well-documented and focused on ionospheric research, not mind control. The scientific community has consistently found no evidence to support these theories, highlighting the importance of scientific literacy in evaluating such claims. Through rigorous research and expert analysis, it is evident that the mind control narrative is a product of misunderstanding and misinformation rather than scientific reality.

Chapter 6: Media and Public Perception

The Role of Traditional Media

The role of traditional media in shaping public perception of HAARP has been significant, influencing how the facility is viewed through both sensationalist and investigative lenses. Traditional media outlets, including newspapers, television, and radio, have each contributed to the complex narrative surrounding HAARP, either fueling conspiracy theories or attempting to debunk them.

Newspaper Coverage

Newspapers, both local and national, have played a crucial role in disseminating information about HAARP. Local newspapers near Gakona, Alaska, where HAARP is located, often provided straightforward coverage of the facility's activities, focusing on its scientific missions and community interactions. In contrast, national newspapers tended to sensationalize HAARP, especially in articles that explored the more controversial and conspiratorial aspects of the program. For instance, a headline in a national tabloid might scream "Secret Government Project Controls Weather," attracting readers with a mix of intrigue and fear. This sensationalist approach often overshadowed the more mundane scientific explanations, leaving the public with skewed perceptions.

Television and Radio

Television and radio have been powerful mediums in spreading information about HAARP. Investigative journalism programs on television have occasionally taken a serious look at HAARP, interviewing scientists and presenting detailed explanations of its research. However, these efforts have often been overshadowed by more sensationalist programming. Shows like "Conspiracy Theories with Jesse Ventura" brought HAARP into the limelight, but not always in the most factual manner. Ventura's program, for example, featured episodes that suggested HAARP's involvement in weather manipulation and mind control, using dramatic visuals and speculative narration to engage viewers. The impact of such programs cannot be underestimated, as they reach large audiences and can significantly influence public opinion.

Radio shows, particularly those dedicated to fringe topics and conspiracy theories, have also played a significant role. Programs like "Coast to Coast AM," which attracts listeners interested in the paranormal and the unexplained, frequently discuss HAARP. Hosts and guests on these shows often speculate about HAARP's capabilities, linking it to various natural disasters and global events. The reach of radio, especially late-night talk shows, means that these theories can spread quickly and become ingrained in the public consciousness.

Impact on Public Opinion

The influence of traditional media on public opinion about HAARP is profound. Audience demographics play a significant role in shaping perceptions. Older generations, who are more likely to consume traditional media such as newspapers and television, may have different views on HAARP compared to younger individuals who rely more on digital platforms. Sensationalist headlines and dramatic television segments can leave lasting impressions, especially when there is a lack of balanced reporting to counter the more extreme claims.

The role of traditional media in shaping the narrative around HAARP highlights the need for responsible journalism. Media outlets have the power to either demystify scientific projects through thorough and balanced reporting or to perpetuate myths and fears through sensationalism. The public's understanding of HAARP, like many other scientific endeavors, is heavily influenced by how it is portrayed in the media. As such, it is crucial for journalists to strive for accuracy and context in their coverage, providing audiences with the information they need to form well-informed opinions.

In summary, traditional media has played a dual role in shaping public perception of HAARP. While some outlets have attempted to provide balanced and factual reporting, others have leaned into sensationalism, contributing to the spread of conspiracy theories. Understanding this dynamic is key to comprehending the broader narrative that surrounds HAARP and its place in the public imagination.

The Influence of Documentaries and Books

The impact of documentaries and books on the spread of HAARP-related conspiracy theories is significant. These media forms have played a crucial role in shaping public perception by providing detailed narratives that often blend factual information with speculative and sensational claims. This section examines notable documentaries and books that have contributed to the HAARP conspiracy narrative, analyzing their content, reach, and the cross-media impact they have generated.

Notable Documentaries

Several documentaries have delved into the mysteries and controversies surrounding HAARP, with varying degrees of credibility and sensationalism. One such documentary is *"Angels Still Don't Play This HAARP,"* which builds on the theories presented in the book *"Angels Don't Play This HAARP"* by Nick Begich and Jeane Manning. This film presents HAARP as a covert weapon capable of weather manipulation and mind control. It uses dramatic visuals, expert interviews, and speculative narration to engage viewers and suggest that HAARP's true purpose is being hidden from the public.

Another influential documentary is *"The Conspiracy Show with Richard Syrett,"* which features an episode on HAARP. Syrett's program, known for exploring various conspiracy theories, presents HAARP in a similar light, questioning its scientific missions and suggesting that it is a tool for government control. These documentaries, through their compelling storytelling and high production values, have reached wide audiences, perpetuating the myths and suspicions surrounding HAARP.

The narrative techniques used in these documentaries often include a mix of factual data, anecdotal evidence, and speculative assertions. By framing the information in a way that suggests a hidden truth, these documentaries can make the conspiracy theories seem plausible, even to skeptical viewers. The reception of these documentaries has been mixed, with critics pointing out their lack of scientific rigor while audiences intrigued by the possibilities they present continue to spread these ideas.

Influential Books

Books have also played a crucial role in propagating HAARP conspiracy theories. One of the most significant works is *"Angels Don't Play This HAARP,"* co-authored by Nick Begich and Jeane Manning. This book is often cited as a foundational text for those who believe in the more sinister applications of HAARP. Begich and Manning combine scientific references with speculative theories, suggesting that HAARP could be used for weather control, mind control, and other nefarious purposes. The book's accessibility and detailed exploration of the subject have made it a key resource for conspiracy theorists.

Other influential books include *"The Pentagon's Brain"* by Annie Jacobsen, which, while not solely focused on HAARP, discusses various government projects, including HAARP, in the context of military and defense research. Jacobsen's investigative approach provides a more balanced view but still raises questions about the potential misuse of such technologies, further feeding public intrigue and suspicion.

The backgrounds and motivations of the authors of these books vary. Nick Begich, for instance, has a background in politics and activism, and his work on HAARP reflects his broader concerns about government transparency and public accountability. Jeane Manning, a journalist, has focused on alternative energy and environmental issues, bringing her investigative skills to the exploration of HAARP.

Cross-Media Impact

The influence of these documentaries and books extends beyond their immediate audiences, affecting other forms of media coverage. News articles, blog posts, and social media content often cite these works, amplifying their reach. Authors and filmmakers frequently appear on television and radio programs to discuss their findings, further embedding HAARP conspiracy theories in the public consciousness.

In summary, documentaries and books have significantly shaped public perception of HAARP by providing detailed and compelling narratives that blend fact with speculation. These media forms have not only reached wide audiences but have also influenced other media coverage, perpetuating the myths and conspiracy theories surrounding HAARP. Understanding this influence is key to addressing the spread of misinformation and promoting a more accurate understanding of HAARP and its scientific missions.

The Role of Online Content and Social Media

In the digital age, the influence of online content and social media on public perception cannot be overstated. These platforms have become primary sources of information and misinformation about HAARP, significantly impacting how the public understands and engages with the facility. This section examines the role of dedicated websites, blogs, social media platforms, and the creation of echo chambers in shaping the narrative around HAARP.

Websites and Blogs

Dedicated websites and blogs have been instrumental in spreading both information and misinformation about HAARP. These platforms often present themselves as alternative news sources, offering explanations that mainstream media allegedly overlooks or suppresses. Websites like Infowars, operated by conspiracy theorist Alex Jones, have featured extensive content on HAARP, linking it to various natural disasters and government plots. These sites often blend factual information about HAARP's capabilities with speculative theories, creating a compelling but misleading narrative.

The content on these websites typically lacks the rigorous scrutiny of peer-reviewed scientific research, instead relying on anecdotal evidence and speculative connections. Blog posts and articles often sensationalize HAARP's activities, suggesting that the facility's research into the ionosphere is a cover for more sinister operations. By presenting these theories as credible alternatives to mainstream explanations, these sites contribute to a broader culture of suspicion and mistrust.

Social Media Platforms

Social media platforms like Facebook, Twitter, and YouTube have amplified the reach of HAARP-related content, making it easy for conspiracy theories to spread rapidly. These platforms utilize algorithms that prioritize engaging and sensational content, often leading to the viral spread of misinformation. For example, a single tweet or Facebook post linking HAARP to a recent natural disaster can be shared and reshared thousands of times, reaching a vast audience in a matter of hours.

YouTube, in particular, has played a significant role in the dissemination of HAARP conspiracy theories. Videos that explore and speculate about HAARP's capabilities often attract millions of views. Channels dedicated to conspiracy theories create video content that presents HAARP as a secretive and powerful entity capable of manipulating weather and minds. These videos use dramatic visuals, ominous music, and persuasive narration to draw viewers in and convince them of the theories' validity.

Creation of Echo Chambers

One of the most significant impacts of online content and social media is the creation of echo chambers, where users are exposed primarily to information that reinforces their existing beliefs. Algorithms on platforms like Facebook and YouTube are designed to show users content that aligns with their interests and viewing history. As a result, someone who watches a HAARP conspiracy video is likely to be recommended more similar content, creating a feedback loop that reinforces their beliefs.

Echo chambers can have profound psychological effects, including confirmation bias, where individuals favor information that confirms their preconceptions and ignore contradictory evidence. This environment makes it difficult for scientific rebuttals and factual information to penetrate, as users are constantly bombarded with content that supports the conspiracy theories.

Efforts to counteract misinformation and promote accurate information about HAARP on these platforms have included fact-checking initiatives and the promotion of credible scientific sources. Social media companies have also taken steps to reduce the spread of misinformation by flagging or removing content that violates their policies. However, the sheer volume of content and the sophisticated tactics used by conspiracy theorists to evade detection present ongoing challenges.

In summary, online content and social media have played a pivotal role in shaping public perception of HAARP. Through dedicated websites, blogs, and social media platforms, conspiracy theories about HAARP have reached vast audiences and created echo chambers that reinforce these beliefs. Understanding the dynamics of these digital environments is crucial for addressing the spread of misinformation and fostering a more informed public discourse.

Chapter 7: Scientific Community's Response

Statements from Researchers

The response from the scientific community regarding the myriad conspiracy theories about HAARP has been clear, consistent, and grounded in empirical evidence. Scientists and researchers directly involved with HAARP, as well as those in related fields, have provided numerous statements to clarify the facility's true purpose and capabilities, dispelling the myths that have surrounded it.

Direct Involvement

Researchers who have worked directly with HAARP are among the most credible voices in debunking the conspiracy theories. Dr. John Heckscher, a former program manager at HAARP, has been vocal in his attempts to demystify the facility's operations. Heckscher has repeatedly emphasized that HAARP's primary mission is to study the ionosphere—a layer of Earth's atmosphere that is vital for radio communications and navigation. In interviews and public statements, Heckscher has explained that HAARP's high-frequency (HF) transmitters are used to induce small, controlled disturbances in the ionosphere. These disturbances allow scientists to study ionospheric behaviors and their effects on radio wave propagation.

Scientists from the University of Alaska Fairbanks (UAF), which now operates HAARP, have also been active in addressing misconceptions. Dr. Chris Fallen, a UAF research assistant professor who has conducted numerous experiments at HAARP, has stated that the facility's work is purely scientific. He has highlighted how HAARP experiments contribute to understanding natural phenomena like the auroras, improve satellite communications, and even provide insights into mitigating the effects of space weather on power grids and communication systems.

Clarification of Objectives

The primary scientific objectives of HAARP have been clearly outlined by researchers to counter the sensationalist claims. HAARP's goals include advancing knowledge about the ionosphere, enhancing the reliability of satellite and radio communications, and understanding the natural processes that affect these technologies. By studying the ionosphere, scientists aim to develop models that can predict ionospheric conditions, which is crucial for improving global positioning systems (GPS), over-the-horizon radar, and other communication technologies.

Researchers have consistently refuted specific conspiracy theories, such as the claims that HAARP can control the weather or manipulate minds. Dr. Heckscher and his colleagues have pointed out that the energy levels generated by HAARP are far too low to influence weather systems or human thought processes. They have emphasized that the ionospheric research conducted at HAARP involves energy outputs that are negligible compared to the energy required to produce any significant atmospheric changes or impact human cognition.

Public Outreach Efforts

In addition to providing statements, scientists have engaged in various public outreach efforts to educate the public about HAARP's true purpose. These efforts include public talks, interviews, and educational materials aimed at explaining the science behind HAARP in accessible terms. For example, the Geophysical Institute at UAF has hosted open houses at HAARP, inviting the public to tour the facility, meet the researchers, and learn about ongoing experiments. These events are designed to promote transparency and build trust by providing firsthand insights into HAARP's operations.

Publications and media appearances by HAARP researchers also serve as valuable tools for demystifying the facility. Articles in scientific journals, popular science magazines, and interviews on reputable news outlets help reach a broader audience. By consistently presenting accurate information and addressing misconceptions directly, scientists aim to foster a better understanding of HAARP's legitimate scientific contributions and dispel the unfounded conspiracy theories.

In summary, the scientific community's response to HAARP-related conspiracy theories has been grounded in clarity, transparency, and evidence-based communication. Researchers involved with HAARP and related fields have provided detailed explanations of the facility's objectives, capabilities, and the nature of its experiments, countering the myths with factual information and public engagement.

Published Studies

The scientific community's evaluation of HAARP's activities has produced a substantial body of peer-reviewed studies and scientific papers that unequivocally demonstrate the facility's true capabilities and limitations. These studies provide critical insights into the objectives and outcomes of HAARP's research, countering the many conspiracy theories that surround the facility.

Overview of Studies

A review of the scientific literature on HAARP reveals a focus on understanding the ionosphere and its interactions with radio waves. One significant study, published in the *Journal of Geophysical Research*, detailed experiments conducted at HAARP to investigate the generation of artificial auroras. The researchers used HAARP's high-frequency transmitters to stimulate the ionosphere, creating visible auroras that were then studied to better understand the natural auroral processes. This study, like many others, emphasized HAARP's role in advancing scientific knowledge about the ionosphere rather than manipulating the environment or human behavior.

Another critical study, featured in *Radio Science*, examined the effects of high-frequency wave modulation on ionospheric plasma density. The findings highlighted how controlled experiments at HAARP could generate plasma waves and small-scale ionospheric irregularities, which are essential for improving the accuracy and reliability of radio communication and navigation systems. These studies consistently reflect HAARP's focus on scientific exploration and technological advancement, without any indication of the more nefarious capabilities suggested by conspiracy theorists.

Key Findings

The key findings from these studies reinforce the primary scientific objectives of HAARP. For example, research has shown that HAARP's ionospheric heating experiments can create temporary, localized changes in the ionosphere, which help scientists develop models to predict space weather and its impacts on Earth-based technologies. These models are crucial for mitigating the effects of geomagnetic storms on power grids, satellite operations, and communication networks.

Studies have also demonstrated HAARP's contributions to understanding natural phenomena such as the Northern Lights. By artificially inducing auroral displays, researchers can gather data in a controlled setting, leading to a deeper understanding of the mechanisms driving these spectacular natural events. This research not only satisfies scientific curiosity but also has practical applications, such as improving the prediction of space weather events that can affect global communication and navigation systems.

Importantly, these studies have consistently found that the energy levels and frequencies used in HAARP's experiments are insufficient to influence large-scale weather patterns or human cognition. This directly counters the conspiracy theories that claim HAARP can control the weather or manipulate minds. The peer-reviewed evidence supports the conclusion that HAARP's activities are strictly within the bounds of ionospheric research and do not extend to environmental or psychological manipulation.

Rebuttal of Theories

The scientific literature provides a robust rebuttal to the various conspiracy theories about HAARP. For instance, a comprehensive review by the American Geophysical Union concluded that HAARP's high-frequency transmissions do not have the power to affect weather systems. The review emphasized that the energy output of HAARP is several orders of magnitude lower than what would be required to influence atmospheric conditions on a significant scale.

Similarly, studies on the biological effects of electromagnetic fields (EMFs) have shown that the frequencies used by HAARP are far too low to impact human health. Research published in *Health Physics* examined the potential health risks of EMF exposure and found no evidence to support claims that low-level electromagnetic fields, such as those produced by HAARP, can affect brain function or behavior. These findings are corroborated by numerous other studies that have investigated the safety of EMF exposure.

In conclusion, the published studies and peer-reviewed research on HAARP provide a clear and evidence-based understanding of its capabilities and limitations. These studies debunk the conspiracy theories by demonstrating that HAARP's activities are focused on scientific inquiry and technological improvement, without any hidden agendas or nefarious purposes. The scientific community's rigorous evaluation of HAARP reinforces the importance of relying on empirical evidence to understand complex scientific endeavors.

General Consensus within the Scientific Field

The scientific community has reached a broad consensus regarding HAARP's activities and capabilities, reinforcing the notion that the facility's research is rooted in legitimate scientific inquiry rather than the nefarious purposes suggested by conspiracy theories. This consensus has significant implications for public perception, policy decisions, and the broader role of scientific organizations in supporting research endeavors like HAARP.

Scientific Consensus

The general agreement among scientists about HAARP's purpose and activities is built on a foundation of transparent research and peer-reviewed studies. The consensus underscores that HAARP is designed to study the ionosphere, a critical component of the Earth's upper atmosphere, to enhance our understanding of its dynamics and to improve communication and navigation technologies. This understanding is essential for mitigating the effects of space weather on satellite systems, power grids, and global communication networks.

Prominent scientific organizations, such as the American Geophysical Union and the Institute of Electrical and Electronics Engineers (IEEE), have reviewed HAARP's research outputs and methodologies. Their evaluations consistently affirm that HAARP's experiments are scientifically sound and pose no threat of environmental or psychological manipulation. These organizations highlight the facility's contributions to advancing ionospheric science and underscore the importance of such research for technological and societal benefits.

Broader Implications

The consensus within the scientific community influences policy decisions and funding for projects like HAARP. Policymakers rely on scientific expertise to guide investments in research and development, ensuring that resources are allocated to projects with demonstrable benefits. The scientific consensus around HAARP has helped secure funding and support for its continued operation, emphasizing its value in advancing knowledge and innovation.

The support from scientific institutions also plays a critical role in maintaining the integrity and credibility of HAARP's research. By aligning HAARP's objectives with broader scientific goals, these institutions help legitimize the facility's work and counteract the misinformation propagated by conspiracy theories. This alignment is crucial for fostering public trust in scientific research and ensuring that legitimate scientific endeavors are not undermined by unfounded fears.

Public Perception and Trust

The scientific community's efforts to build and maintain public trust are vital in shaping the public's perception of HAARP. Transparency and open communication are key strategies in this effort.

Researchers and institutions involved with HAARP have made concerted efforts to engage with the public through open houses, educational programs, and media outreach. These initiatives aim to demystify HAARP's activities and provide clear, evidence-based explanations of its research.

Despite these efforts, public perception is often influenced by sensationalist media coverage and conspiracy theories. However, the consistent message from the scientific community helps counteract these influences. By providing accurate information and addressing misconceptions directly, scientists work to ensure that the public has a reliable understanding of HAARP's true purpose and capabilities.

Efforts to promote scientific literacy also play a crucial role in shaping public perception. Educational programs that emphasize critical thinking and the scientific method help individuals evaluate claims about HAARP and other scientific projects more effectively. By fostering a more informed and discerning public, the scientific community can mitigate the impact of misinformation and conspiracy theories.

In conclusion, the general consensus within the scientific community regarding HAARP is clear and robust. This consensus, built on empirical evidence and peer-reviewed research, emphasizes HAARP's legitimate scientific objectives and refutes the sensational claims made by conspiracy theorists. The broader implications of this consensus for policy, public perception, and scientific integrity underscore the importance of transparent communication and public engagement in maintaining trust in scientific research. Through these efforts, the scientific community continues to support and validate HAARP's contributions to our understanding of the ionosphere and its impact on global communication and navigation systems.

Chapter 8: Government and Military Involvement

Historical Context and Initial Development

The inception of the High-Frequency Active Auroral Research Program (HAARP) in the early 1990s was a product of its time—an era defined by rapid technological advancements and shifting geopolitical landscapes. Understanding the historical context and the motivations behind HAARP's creation requires a look at the geopolitical and technological factors that influenced its development, as well as the roles played by key government and military entities.

Origins and Motivations

The early 1990s marked the end of the Cold War, a period characterized by intense rivalry between the United States and the Soviet Union. The dissolution of the Soviet Union in 1991 shifted the global power dynamics and prompted the United States to reassess its national security strategies and technological capabilities. During this time, the U.S. government sought to advance its understanding of ionospheric science to improve communication and surveillance systems, which were critical for both military and civilian applications.

HAARP was conceived as a research initiative aimed at exploring the ionosphere—a layer of the Earth's atmosphere that plays a crucial role in radio wave propagation. The project's primary objectives included studying ionospheric interactions, enhancing over-the-horizon radar systems, and improving satellite communications and navigation. The need for advanced ionospheric research was driven by the desire to secure reliable and effective communication channels, which were vital for national defense and global operations.

Project Development

The development of HAARP involved multiple government agencies and military branches. The U.S. Air Force and the U.S. Navy were the primary sponsors of the project, with additional support from the Defense Advanced Research Projects Agency (DARPA). These agencies recognized the strategic importance of ionospheric research and committed resources to the construction and operation of HAARP.

The planning and design of HAARP began in the early 1990s, with the goal of creating a state-of-the-art research facility in Gakona, Alaska. The remote location was chosen for its unique geographical advantages, including minimal electromagnetic interference and proximity to the auroral zone, which made it an ideal site for ionospheric studies. Construction of the facility commenced in 1993, and HAARP became operational in 1995.

The project's development saw the collaboration of various military branches, government agencies, and academic institutions. Scientists and engineers from the Air Force Research Laboratory (AFRL) and the Naval Research Laboratory (NRL) played key roles in designing the high-frequency transmitters and antenna arrays that form the core of HAARP's capabilities. Additionally, universities and research institutions contributed their expertise in atmospheric science and radio wave propagation, ensuring that HAARP's scientific objectives were integrated with military interests.

Collaboration with Academic Institutions

The collaboration between government, military, and academic entities was a hallmark of HAARP's development. Academic institutions such as the University of Alaska Fairbanks were involved from the outset, providing scientific guidance and conducting joint research projects. This partnership aimed to ensure that HAARP's research not only advanced military and national security interests but also contributed to the broader scientific understanding of the ionosphere.

The integration of scientific and military goals was carefully balanced to maximize the benefits of HAARP's research. While the military's interest lay in enhancing communication and surveillance technologies, the scientific community focused on understanding fundamental ionospheric processes. This dual focus allowed HAARP to serve both practical and academic purposes, driving innovations in technology and expanding the frontiers of atmospheric science.

In summary, HAARP's development was deeply rooted in the geopolitical and technological context of the early 1990s. The involvement of the U.S. government and military was driven by the need to advance national security capabilities and scientific knowledge. Through collaboration with academic institutions, HAARP achieved its dual objectives of enhancing communication technologies and contributing to the understanding of the ionosphere, setting the stage for its continued research endeavors.

Public Concerns and Suspicions

The involvement of the U.S. government and military in the High-Frequency Active Auroral Research Program (HAARP) has been a major source of public concern and suspicion. Understanding these concerns requires an exploration of the factors contributing to public mistrust, the historical precedents that fuel these suspicions, and the specific conspiracy theories that have emerged around HAARP.

Public Mistrust

Public mistrust of government and military projects is not a new phenomenon. Historical precedents of covert operations and secret experiments have left a lingering sense of skepticism and fear among the public. One of the most infamous examples is the CIA's MKUltra program, which conducted unethical experiments on mind control and behavior modification without the knowledge or consent of the subjects involved. Revelations about such projects have eroded public trust and made people wary of other government-funded research initiatives, including HAARP.

The secrecy often associated with military projects also contributes to public suspicion. Many people believe that if a project is shrouded in secrecy, it must be hiding something nefarious. HAARP's remote location in Gakona, Alaska, and its funding by military entities like the U.S. Air Force and Navy have only fueled these fears. For those who already distrust the government, these factors are seen as evidence that HAARP is involved in activities beyond its stated scientific goals.

Conspiracy Theories

Several conspiracy theories have emerged around HAARP, each alleging different hidden agendas and capabilities. One of the most prevalent theories is that HAARP is used for weather control. Conspiracy theorists claim that HAARP's high-frequency transmissions can manipulate the weather, causing natural disasters like hurricanes, droughts, and earthquakes. These theories often point to unusual weather patterns and natural disasters as "evidence" of HAARP's capabilities, despite the lack of scientific support for such claims.

Another popular theory is that HAARP is involved in mind control. This theory suggests that HAARP's radio waves can influence human thoughts and behavior, effectively turning the facility into a tool for mass psychological manipulation. Proponents of this theory often cite the history of government experiments in mind control as a basis for their claims, ignoring the significant scientific evidence debunking such possibilities with HAARP's technology.

Environmental manipulation is another conspiracy theory linked to HAARP. This theory posits that HAARP can alter the environment in ways that could be harmful to ecosystems and human health. For example, some theorists believe that HAARP's transmissions can create artificial auroras that disturb wildlife or generate radiation levels harmful to humans. These claims are typically based on a misunderstanding of HAARP's scientific operations and the nature of its research.

Media Amplification

The role of the media in amplifying these conspiracy theories cannot be overlooked. Sensationalist media coverage often prioritizes dramatic and alarming narratives over factual reporting, which can spread fear and misinformation. Television shows, online articles, and social media platforms frequently feature HAARP in the context of conspiracy theories, giving these ideas a broader audience and more perceived legitimacy.

High-profile figures and incidents have also contributed to the spread of HAARP-related conspiracy theories. For instance, public figures like Alex Jones and Jesse Ventura have used their platforms to promote these theories, reaching millions of people and cementing HAARP's place in the conspiracy theory canon. Additionally, significant natural disasters and unusual weather events are often quickly attributed to HAARP by conspiracy theorists, who use these incidents to validate their claims.

In summary, public concerns and suspicions about HAARP stem from a combination of historical mistrust, the facility's secrecy, and the amplification of conspiracy theories by the media. These factors have created an environment where speculation and fear often overshadow factual information. Addressing these concerns requires transparency, effective communication, and a commitment to debunking misinformation through credible scientific evidence and public engagement.

Transparency and Oversight

Despite the persistent conspiracy theories and public concerns, HAARP has implemented numerous measures to ensure transparency and maintain robust oversight. This section delves into the regulatory frameworks governing HAARP's operations, the transparency initiatives undertaken to inform and engage the public, and how these efforts have helped address and alleviate public concerns.

Oversight and Regulation

HAARP operates under stringent regulatory frameworks and oversight mechanisms designed to ensure its activities are conducted responsibly and transparently. The facility is subject to oversight by multiple governmental bodies, including the Federal Communications Commission (FCC), which regulates the use of radio frequencies, and the Environmental Protection Agency (EPA), which monitors environmental impacts. Additionally, the Department of Defense (DoD) oversees HAARP's operations through its branches, including the Air Force and Navy, ensuring compliance with national security protocols and ethical standards.

The process of approving and reviewing research projects at HAARP involves a rigorous evaluation to ensure they align with the facility's scientific and technological goals. Each project proposal undergoes scrutiny from a panel of experts, including representatives from academic institutions and government agencies. This collaborative approach ensures that research conducted at HAARP is scientifically sound and ethically conducted, reinforcing the facility's commitment to transparency and accountability.

Transparency Initiatives

To combat misinformation and foster public trust, HAARP has undertaken several transparency initiatives aimed at demystifying its operations. One such effort is the annual open house event, where the public is invited to tour the facility, interact with researchers, and learn about ongoing experiments. These events provide a unique opportunity for individuals to see firsthand the scientific work being conducted and to ask questions directly to the scientists involved. By opening its doors to the public, HAARP demonstrates its commitment to transparency and dispels some of the mystery surrounding its activities.

Educational initiatives have also been a cornerstone of HAARP's transparency efforts. The facility collaborates with academic institutions to develop educational programs and materials that explain the science behind ionospheric research. These programs are designed to be accessible to a wide audience, from students to the general public, and aim to enhance scientific literacy and understanding of HAARP's objectives. Additionally, HAARP's research findings are frequently published in peer-reviewed journals and presented at scientific conferences, further promoting openness and scholarly exchange.

Addressing Public Concerns

HAARP's operators have actively sought to address public concerns and criticisms through open communication and responsive policies. When faced with specific allegations or fears, HAARP representatives have provided detailed explanations and evidence to counter these claims. For example, in response to concerns about potential environmental impacts, HAARP has conducted and published environmental impact assessments, demonstrating that its operations do not harm local ecosystems or human health.

Efforts to engage with the media have also been critical in addressing public concerns. HAARP researchers and spokespeople have participated in interviews, documentaries, and public forums to explain their work and refute misconceptions. These engagements help to correct misinformation and provide the public with a more accurate understanding of HAARP's mission and capabilities.

Continuous dialogue between HAARP's researchers, government agencies, and the public is essential for maintaining trust and addressing ongoing concerns. By fostering an open exchange of information and being responsive to public inquiries, HAARP aims to build a more informed and supportive community. This proactive approach not only helps mitigate the spread of conspiracy theories but also highlights the positive impacts of HAARP's research on scientific advancement and technological innovation.

In conclusion, HAARP has implemented comprehensive measures to ensure transparency and robust oversight of its operations. Through regulatory compliance, public outreach, and educational initiatives, HAARP demonstrates its commitment to ethical research and public engagement. These efforts have been instrumental in addressing public concerns, dispelling myths, and fostering trust in the scientific community. By continuing to prioritize transparency and open communication, HAARP aims to maintain its role as a leading institution in ionospheric research and technological development.

Chapter 9: Case Studies of Alleged HAARP Interfere

Unusual Weather Patterns

The High-Frequency Active Auroral Research Program (HAARP) has been at the center of numerous conspiracy theories, particularly those suggesting that the facility has the capability to influence weather patterns. This section examines several high-profile weather events that have been attributed to HAARP, assessing the validity of these claims through scientific investigation and expert analysis.

Hurricane Katrina (2005)

One of the most devastating natural disasters in recent history, Hurricane Katrina, has been linked by conspiracy theorists to HAARP. The claim posits that HAARP's high-frequency transmissions influenced the formation and path of the hurricane, intensifying its destructive power. Proponents of this theory argue that the timing and intensity of the storm were unnatural and that HAARP's capabilities were employed to manipulate the weather.

To assess these claims, it is essential to examine the meteorological data and expert analysis surrounding Hurricane Katrina. The hurricane formed in the Bahamas in August 2005 and rapidly intensified as it moved through the warm waters of the Gulf of Mexico. Meteorologists have identified several natural atmospheric conditions that contributed to its strength, including unusually warm sea surface temperatures, low wind shear, and high humidity levels. These factors created a conducive environment for the hurricane's development and intensification.

Scientific experts, including those from the National Oceanic and Atmospheric Administration (NOAA), have extensively studied Hurricane Katrina and found no evidence to support the notion that HAARP influenced its formation or path. The energy required to significantly alter weather patterns on such a scale far exceeds the capabilities of HAARP's transmitters, which are designed for localized ionospheric research rather than large-scale atmospheric manipulation.

Russian Heatwave (2010)

Another significant weather event often linked to HAARP is the 2010 heatwave in Russia. This extreme weather event resulted in record-high temperatures, severe drought, and devastating wildfires, causing widespread damage and loss of life. Conspiracy theorists have suggested that HAARP was responsible for this heatwave, arguing that its transmissions could manipulate regional climate conditions to create such anomalies.

To evaluate these claims, it is necessary to analyze the temperature records and atmospheric conditions of the time. The heatwave was characterized by a persistent high-pressure system, known as a blocking high, which trapped hot air over Russia for an extended period. Climate scientists have explained that such blocking patterns are natural phenomena that can occur due to variations in the jet stream and other atmospheric dynamics.

Temperature records from the Russian heatwave indicate that the extreme heat was part of a broader pattern of weather anomalies observed across Europe and Asia during that summer. These patterns are consistent with natural climate variability and are not indicative of artificial manipulation. Scientific studies, including those conducted by the World Meteorological Organization (WMO), have concluded that the heatwave was a result of natural atmospheric processes rather than human-induced interference from HAARP.

Unseasonal Storms

Conspiracy theorists have also attributed various unseasonal storms to HAARP, suggesting that the facility has the capability to generate or intensify storms in regions where they are not typically expected. Examples include winter storms occurring in regions usually experiencing mild weather or sudden severe weather events in typically calm areas.

To debunk these theories, it is important to consider the scientific principles of weather formation. Storms are complex systems influenced by a multitude of factors, including temperature gradients, humidity, and atmospheric pressure. Meteorologists use advanced models to predict storm patterns, which are driven by natural climate variability and regional weather conditions.

In specific cases of unseasonal storms, scientific evidence points to natural causes such as shifts in the jet stream, El Niño or La Niña events, and other large-scale climatic patterns. These factors can create unusual weather conditions, but there is no scientific basis for attributing them to HAARP's activities. Expert commentary from meteorologists consistently reinforces that the energy output of HAARP is insufficient to influence weather patterns on a significant scale.

In conclusion, the scientific investigation of these unusual weather patterns reveals that the claims of HAARP's influence lack credible evidence. Natural atmospheric conditions and climate variability provide sufficient explanations for these events, disproving the need for theories involving artificial manipulation. By understanding the science behind weather phenomena, we can dispel the myths and focus on the real factors that shape our climate.

Natural Disasters

HAARP has been implicated by conspiracy theorists in various natural disasters, from earthquakes to tsunamis, despite the lack of scientific evidence supporting these claims. This section investigates several high-profile natural disasters purportedly influenced by HAARP, using geological data and scientific analysis to assess the validity of these assertions.

Haiti Earthquake (2010)

The devastating earthquake that struck Haiti on January 12, 2010, has been one of the focal points for conspiracy theories involving HAARP. Some theorists claim that HAARP's high-frequency transmissions could have triggered the earthquake, citing the timing and magnitude of the event as evidence. However, a thorough examination of geological data and expert analysis disproves these theories.

The Haiti earthquake, which registered a magnitude of 7.0, occurred along the Enriquillo-Plantain Garden fault, a significant strike-slip fault system in the region. Seismologists have long studied this fault, which has a history of seismic activity due to the tectonic movement between the Caribbean and North American plates. The build-up of stress along this fault line eventually led to a sudden release of energy, resulting in the earthquake.

Geological data, including seismographic readings, confirm that the earthquake's origin was deep within the Earth's crust, where HAARP's influence could not possibly reach. The energy required to cause such a seismic event is far beyond the capabilities of HAARP's transmitters, which are designed for ionospheric research and not for affecting geological processes. The scientific consensus is clear: the Haiti earthquake was a natural geological event, and there is no credible evidence linking it to HAARP.

Japanese Tōhoku Earthquake and Tsunami (2011)

Another major natural disaster frequently associated with HAARP by conspiracy theorists is the Tōhoku earthquake and tsunami that struck Japan on March 11, 2011. This catastrophic event caused widespread devastation, including the Fukushima nuclear disaster. Theorists allege that HAARP's transmissions could have induced the earthquake and subsequent tsunami, a claim that lacks scientific support.

The Tōhoku earthquake, with a magnitude of 9.0, was one of the most powerful earthquakes ever recorded. It occurred in a subduction zone where the Pacific Plate is being forced beneath the North American Plate. This tectonic setting is known for generating large, destructive earthquakes due to the immense pressure and stress accumulated along the plate boundary. The sudden release of this stress caused the earthquake and triggered the massive tsunami.

Geological and oceanographic data provide a comprehensive understanding of the Tōhoku earthquake's natural causes. Detailed seismographic analyses show that the earthquake originated deep beneath the ocean floor, far beyond the reach of HAARP's high-frequency emissions. Furthermore, the energy involved in tectonic movements at such depths is orders of magnitude greater than anything HAARP could produce. The scientific community, including experts from the Japan Meteorological Agency and the U.S. Geological Survey, unanimously agrees that the Tōhoku earthquake was a result of natural tectonic processes.

Volcanic Eruptions

Some conspiracy theories even extend to volcanic eruptions, attributing events like the Eyjafjallajökull eruption in Iceland (2010) to HAARP's activities. The eruption caused significant disruption to air travel across Europe and has been linked by some to HAARP, claiming that the facility's emissions could have triggered the volcanic activity.

To assess these claims, it is necessary to understand the geological processes underlying volcanic eruptions. Volcanic activity is driven by the movement of magma within the Earth's mantle and crust, caused by the intense heat and pressure from the Earth's interior. The Eyjafjallajökull eruption, like other volcanic events, was the result of natural geothermal processes.

Geological evidence and expert analysis from volcanologists demonstrate that HAARP lacks the capability to influence volcanic activity. The energy and mechanisms required to trigger an eruption are vastly different from the ionospheric research conducted at HAARP. Volcanic eruptions are complex natural events that occur due to the Earth's internal dynamics, not external electromagnetic influences.

In conclusion, the scientific investigation of these natural disasters reveals that claims of HAARP's involvement are unfounded. Geological and oceanographic data consistently show that these events are the result of natural processes, driven by tectonic and geothermal forces far beyond the reach of HAARP's capabilities. By understanding the science behind these disasters, we can dispel the myths and focus on preparing for and mitigating the impacts of such events in the future.

Psychological Phenomena

The notion that HAARP has the capability to influence human psychology, causing phenomena such as mass anxiety, mind control, and behavioral changes, is a persistent theme among conspiracy theorists. This section examines these claims through detailed case studies, evaluating them using psychological and neuroscientific evidence.

Mass Anxiety Incidents

Mass anxiety incidents attributed to HAARP typically involve sudden and unexplained episodes of fear or panic within communities. Conspiracy theorists argue that HAARP's high-frequency transmissions can trigger widespread anxiety by interfering with brainwave patterns. To evaluate these claims, it is crucial to consider both the psychological and social factors contributing to such incidents.

One notable case often cited by conspiracy theorists is the wave of anxiety and panic attacks reported in the city of Havana, Cuba, in late 2016. Residents experienced symptoms such as dizziness, headaches, and feelings of intense fear, leading some to speculate that HAARP's technology was behind these experiences. However, psychological analysis and expert investigations revealed a different picture.

Social psychologists and neurologists studied the incident and concluded that the symptoms were more likely caused by a combination of stress, environmental factors, and the spread of rumors. The political climate, economic hardships, and heightened tensions in the region created a fertile ground for anxiety. Once a few cases of panic were reported, a form of mass hysteria ensued, where symptoms spread through suggestion and social contagion. These findings underscore that while HAARP makes for a convenient scapegoat, the reality often lies in more mundane psychological and social dynamics.

Mind Control Allegations

The most sensational and widely discussed psychological phenomenon attributed to HAARP is the alleged capability of mind control. Proponents of this theory claim that HAARP's electromagnetic waves can influence thought patterns, manipulate emotions, and even control human behavior. To assess these allegations, it is essential to delve into the scientific principles of brain function and electromagnetic effects.

Conspiracy theorists often cite the history of government experiments in mind control, such as the CIA's MKUltra program, to lend credibility to their claims about HAARP. However, the technology and methods used in MKUltra were vastly different and involved unethical, direct interventions like the administration of psychoactive drugs and psychological torture. HAARP's ionospheric research, on the other hand, involves high-frequency radio waves that interact with the upper atmosphere, far removed from the mechanisms required to influence human cognition.

Neuroscientific studies have shown that brain function is governed by complex electrochemical processes that are not easily disrupted by external electromagnetic fields at the levels generated by HAARP. The facility's transmissions are designed to interact with the ionosphere, several kilometers above the Earth's surface, and lack the precision and intensity needed to affect human neural activity. Experts in neuroscience and psychology consistently refute the feasibility of using HAARP for mind control, emphasizing the distinction between scientifically grounded research and speculative fiction.

Behavioral Changes

Another area of concern is the claim that HAARP can induce widespread behavioral changes in populations. This theory suggests that HAARP's emissions can manipulate mood, behavior, and even public opinion by targeting specific frequencies. To investigate these claims, it is important to review sociological studies and scientific research on behavior modification.

Behavioral changes in populations are influenced by a myriad of factors, including economic conditions, social dynamics, and environmental stressors. For example, economic downturns, political unrest, and social media influence can all contribute to shifts in public behavior and sentiment. These factors are well-documented and provide a more plausible explanation for observed changes than the speculative influence of HAARP.

Scientific research on the effects of electromagnetic fields on human behavior has found no evidence to support the idea that low-level EMFs, such as those produced by HAARP, can alter behavior. The energy levels and frequencies involved are insufficient to cause significant biological effects. Sociologists and psychologists emphasize that attributing behavioral changes to HAARP overlooks the complex interplay of social, economic, and environmental factors that shape human behavior.

In conclusion, the scientific investigation of psychological phenomena purportedly influenced by HAARP reveals that these claims are unsupported by credible evidence. Mass anxiety, mind control, and behavioral changes are complex issues best explained by psychological, social, and environmental factors rather than the speculative capabilities of HAARP. By understanding the science behind these phenomena, we can dispel the myths and focus on the real influences affecting human psychology and behavior.

Chapter 10: Debunking the Myths

Scientific Facts and Logical Reasoning

The myths surrounding the High-Frequency Active Auroral Research Program (HAARP) are numerous and persistent, often driven by misunderstanding and fear of the unknown. To effectively debunk these myths, it is crucial to present the foundational scientific facts about HAARP and use logical reasoning to dismantle the false narratives that have taken hold. Emphasizing evidence-based understanding and critical thinking is essential in this endeavor.

HAARP's Purpose and Capabilities

HAARP was established primarily as an ionospheric research facility. The ionosphere, a region of Earth's upper atmosphere, plays a critical role in radio communication, satellite operation, and navigation systems by reflecting and modifying radio waves. Understanding these processes helps improve the reliability and efficiency of global communication networks.

The core technology at HAARP involves high-frequency (HF) transmitters that emit radio waves aimed at small, localized regions of the ionosphere. These emissions create controlled disturbances, allowing scientists to study ionospheric behavior under various conditions. The scientific goals of HAARP include improving GPS accuracy, enhancing satellite communication, and understanding space weather phenomena, such as solar flares and geomagnetic storms.

Energy Output and Limitations

One of the most significant misconceptions about HAARP is its alleged capability to influence large-scale natural phenomena, such as weather patterns, earthquakes, and even human behavior. To address these claims, it is essential to consider the energy output of HAARP's transmitters in the context of natural processes.

HAARP's transmitters generate high-frequency radio waves with energy levels that are relatively modest compared to the forces driving weather systems and geological activity. For instance, the energy output of a single HAARP transmission is measured in megawatts, which is minuscule compared to the energy released by a typical thunderstorm, which can reach terawatts. Similarly, the energy required to trigger an earthquake or volcanic eruption is orders of magnitude greater than what HAARP can produce.

The scientific principles governing ionospheric research further highlight the limitations of HAARP's capabilities. The facility's experiments are designed to create temporary and localized changes in the ionosphere, primarily for the purpose of studying radio wave propagation and other related phenomena. These changes are not sufficient to influence the broader atmospheric or geological systems, which are governed by complex and powerful natural forces.

Logical Fallacies in Conspiracy Theories

Many of the conspiracy theories surrounding HAARP are based on logical fallacies that undermine their credibility. One common fallacy is the conflation of correlation with causation. For example, the occurrence of a natural disaster following a HAARP transmission might lead some to incorrectly assume a causal link between the two events. However, such assumptions ignore the vast array of natural variables that contribute to these phenomena and the lack of scientific evidence supporting the alleged connection.

Another prevalent fallacy is the appeal to ignorance, which suggests that because something is not fully understood, it must be caused by HAARP. This reasoning exploits gaps in public knowledge about ionospheric science and leverages fear of the unknown. It is crucial to address these gaps with clear, factual information to dispel unfounded fears.

Encouraging critical thinking and skepticism is essential in evaluating extraordinary claims. By questioning the sources of information, evaluating the evidence presented, and recognizing logical fallacies, individuals can develop a more nuanced and accurate understanding of HAARP's true capabilities. Engaging with the scientific community, seeking out peer-reviewed research, and fostering a culture of inquiry and education are key steps in debunking the myths and promoting an evidence-based perspective on HAARP.

In conclusion, presenting the scientific facts about HAARP and using logical reasoning to dismantle the myths is vital for dispelling misinformation. By understanding HAARP's purpose, energy limitations, and the logical fallacies in conspiracy theories, we can foster a more informed and rational discourse about this important scientific facility.

Evidence-Based Arguments

One of the most effective ways to dismantle the myths surrounding HAARP is to present specific evidence and expert testimony that counter the most prevalent conspiracy theories. This section will address the claims of weather control, mind control, and geological manipulation by providing detailed case studies and scientific findings.

Weather Control Claims

Among the most persistent myths about HAARP is the notion that it can control weather patterns. Conspiracy theorists claim that HAARP's high-frequency transmissions can manipulate the atmosphere, leading to the creation or intensification of storms, hurricanes, and other weather phenomena. To debunk this theory, it is essential to present scientific evidence from meteorological studies.

Meteorologists and climate scientists have thoroughly investigated the claims that HAARP can influence weather. One common point of analysis is the energy output of HAARP's transmitters compared to the energy required to drive weather systems. Storms, hurricanes, and other weather events are powered by vast amounts of energy derived from the sun, atmospheric conditions, and oceanic processes. The energy released by a single hurricane, for instance, is equivalent to the energy of several nuclear bombs, far exceeding the capabilities of HAARP's transmitters, which operate on a much smaller scale.

Studies conducted by organizations such as the National Oceanic and Atmospheric Administration (NOAA) have shown that weather patterns are influenced by a complex interplay of natural factors, including sea surface temperatures, atmospheric pressure, and wind patterns. These studies provide robust evidence that weather systems operate independently of any human-made interventions, such as those alleged to be possible with HAARP. Climate experts and meteorologists consistently emphasize that there is no scientific basis for the claim that HAARP can control the weather.

Mind Control Allegations

The theory that HAARP can be used for mind control is another myth that has gained traction among conspiracy theorists. This theory posits that HAARP's electromagnetic waves can influence human brainwaves and manipulate thoughts and behavior. To refute this claim, it is crucial to examine neurological and psychological evidence regarding the effects of electromagnetic fields on the brain.

Neuroscientists and psychologists have conducted extensive research on the interactions between electromagnetic fields (EMFs) and human brain activity. Their findings consistently show that the low-level EMFs produced by HAARP are incapable of affecting brain function. The human brain operates through complex electrochemical processes, with brainwaves occurring at specific frequencies. While certain frequencies of EMFs can affect biological tissues, the levels required to influence brain activity are far higher than those generated by HAARP.

Research published in peer-reviewed journals, such as the *Journal of Neuroscience* and *Health Physics*, supports the conclusion that HAARP's technology cannot impact human thoughts or behavior. Experts in the field, including prominent neuroscientists, have publicly stated that the notion of HAARP being used for mind control is scientifically unfounded. These experts emphasize that brain function is resilient to low-level EMFs, and the idea of using HAARP for psychological manipulation is a misinterpretation of scientific principles.

Geological Manipulation Theories

Conspiracy theories also allege that HAARP can cause earthquakes and volcanic eruptions. These theories suggest that HAARP's transmissions can interfere with tectonic plates, leading to seismic activity. To address these claims, it is important to present geological data and expert analysis showing the natural causes of such phenomena.

Seismologists and geophysicists have extensively studied the causes of earthquakes and volcanic eruptions. These natural events are driven by the movement of tectonic plates and the release of accumulated stress within the Earth's crust. The energy involved in these processes is immense, far exceeding the output of any human-made technology, including HAARP.

Geological data, including seismographic readings and tectonic activity records, consistently show that earthquakes and volcanic eruptions originate from deep within the Earth's crust, where HAARP's high-frequency transmissions cannot reach. Studies from institutions such as the United States Geological Survey (USGS) provide detailed evidence that natural tectonic and geothermal processes are responsible for these events. Seismologists and geophysicists unequivocally state that there is no scientific basis for the claim that HAARP can induce seismic or volcanic activity.

In conclusion, the evidence-based arguments presented by experts in meteorology, neuroscience, and geology comprehensively debunk the myths surrounding HAARP. By understanding the scientific principles and data that govern natural phenomena, we can dismantle these conspiracy theories and promote a more accurate and rational discourse about HAARP and its true capabilities.

Promoting Scientific Literacy and Critical Thinking

Debunking the myths surrounding HAARP requires more than just presenting evidence; it necessitates fostering a culture of scientific literacy and critical thinking. By equipping individuals with the tools to assess information critically and understand basic scientific principles, we can combat misinformation and build a more informed society. This section focuses on the importance of scientific literacy, the development of critical thinking skills, and strategies for combating misinformation.

Importance of Scientific Literacy

Scientific literacy is the ability to understand and apply scientific concepts and processes. It empowers individuals to make informed decisions, evaluate the credibility of information, and engage with scientific issues that affect their lives and society. In the context of HAARP, scientific literacy helps people discern between legitimate scientific research and unfounded conspiracy theories.

Educational systems play a crucial role in fostering scientific literacy. By incorporating comprehensive science education into school curricula, students can develop a strong foundation in key scientific concepts and methodologies. Programs that emphasize hands-on learning, critical analysis, and real-world applications of science can inspire curiosity and a deeper understanding of the natural world.

Public outreach and science communication are also vital components of promoting scientific literacy. Scientists, educators, and institutions must engage with the public through accessible and relatable platforms, such as community talks, social media, and popular science writing. By demystifying complex scientific topics and making them relatable, science communicators can bridge the gap between experts and the general public.

Critical Thinking Skills

Critical thinking is the ability to analyze information, evaluate evidence, and draw reasoned conclusions. It is a fundamental skill for navigating the vast amounts of information available in the digital age and for discerning fact from fiction. Developing critical thinking skills involves questioning assumptions, recognizing biases, and considering multiple perspectives.

One practical approach to fostering critical thinking is through education that encourages inquiry and skepticism. Students should be taught to question sources, seek out evidence, and critically evaluate the reliability and validity of information. Lessons in logical reasoning, argumentation, and the scientific method can help individuals develop a rigorous approach to evaluating claims.

Outside of formal education, critical thinking can be nurtured through everyday practices. For example, individuals can engage with a diverse range of news sources, participate in discussions that challenge their views, and reflect on their reasoning processes. Encouraging a culture of curiosity and skepticism helps build resilience against misinformation and promotes a more informed public discourse.

Combating Misinformation

Misinformation, particularly in the digital age, spreads rapidly and can have significant consequences. Combating misinformation requires a multi-faceted approach that involves scientists, educators, media, and the public. Transparency, open communication, and the promotion of credible sources are key strategies in this effort.

Scientists and experts have a responsibility to communicate their findings clearly and accurately. Engaging with the media, participating in public forums, and leveraging social media platforms are effective ways to reach a broad audience. By providing accurate and timely information, scientists can counteract false narratives and build public trust.

Educational initiatives can also play a significant role in combating misinformation. Media literacy programs that teach individuals how to identify credible sources, fact-check information, and recognize biased reporting are essential. These programs can empower people to navigate the digital information landscape with confidence and discernment.

The role of the media in promoting accurate information is crucial. Journalists and media outlets must adhere to high standards of accuracy and integrity, avoiding sensationalism and verifying facts before publication. Collaborative efforts between media organizations and scientific institutions can enhance the quality and reliability of information disseminated to the public.

Finally, individuals have a role to play in combating misinformation. By engaging in critical thinking, questioning sources, and sharing accurate information, people can contribute to a culture of informed and reasoned discourse. Encouraging open dialogue and providing constructive feedback can help counteract the spread of false information.

In conclusion, promoting scientific literacy and critical thinking is essential for debunking the myths surrounding HAARP and other scientific topics. By fostering a culture of inquiry, skepticism, and evidence-based reasoning, we can combat misinformation and build a more informed and resilient society. Through education, communication, and collaboration, we can ensure that the public is equipped to understand and engage with scientific issues in a meaningful way.

Chapter 11: The Real Science Behind HAARP

Contributions to Atmospheric Science

The High-Frequency Active Auroral Research Program (HAARP) has made significant contributions to the field of atmospheric science, advancing our understanding of the ionosphere and space weather phenomena. This section highlights some of the key areas where HAARP's research has had a profound impact.

Ionospheric Research

HAARP has been instrumental in advancing the study of the ionosphere, particularly in understanding how high-frequency (HF) radio waves interact with this region of the atmosphere. The ionosphere is a dynamic layer filled with charged particles that play a crucial role in radio communications by reflecting and modifying radio waves. By using HF transmitters, HAARP can create controlled disturbances in the ionosphere, allowing scientists to observe and analyze these interactions in detail.

One of the notable achievements of HAARP is its research on ionospheric heating. By directing HF radio waves into the ionosphere, HAARP can heat localized areas, creating artificial plasma clouds. These experiments help scientists understand natural ionospheric processes such as ionospheric turbulence, plasma density variations, and the effects of solar activity on the ionosphere. The ability to generate and study artificial auroras has provided valuable insights into the mechanisms driving natural auroral displays, enhancing our knowledge of space weather phenomena.

Space Weather Monitoring

Another critical area of HAARP's contribution is space weather monitoring and prediction. Space weather refers to the conditions in space influenced by the Sun, such as solar flares, geomagnetic storms, and radiation belts. These phenomena can have significant impacts on satellite operations, communication systems, and navigation technologies.

HAARP's research has improved our ability to monitor and predict space weather events. By studying the ionosphere's response to solar activity, HAARP helps scientists develop models to forecast geomagnetic storms and their potential effects on Earth. Collaborations with other research institutions and space agencies, such as NASA and the European Space Agency (ESA), have enhanced the accuracy and reliability of space weather predictions, benefiting both military and civilian infrastructure.

Practical applications of HAARP's research include improving the resilience of critical infrastructure to space weather events. For example, understanding how geomagnetic storms affect power grids and communication networks allows engineers to design systems that can withstand these disturbances, ensuring continuity of services during space weather events.

Atmospheric Propagation Studies

HAARP has also conducted extensive research on the propagation of radio waves through the atmosphere, with significant implications for communication and radar systems. Understanding how radio waves travel through different layers of the atmosphere is crucial for improving long-range communication, navigation, and surveillance technologies.

Studies at HAARP have focused on how HF radio waves interact with the ionosphere and other atmospheric layers, providing valuable data on signal reflection, refraction, and absorption. These findings have led to advancements in over-the-horizon radar systems, which rely on ionospheric reflection to detect objects beyond the line of sight. Improved knowledge of radio wave propagation has also enhanced the reliability and effectiveness of communication systems used in both civilian and military applications.

For instance, HAARP's research has helped optimize frequencies and transmission techniques for various communication systems, ensuring clearer signals and more robust connections over long distances. This is particularly important for remote and challenging environments where traditional communication infrastructure may be limited.

In summary, HAARP's contributions to atmospheric science are vast and impactful. Through its pioneering research on the ionosphere, space weather monitoring, and atmospheric propagation, HAARP has advanced our understanding of critical atmospheric processes and their implications for technology and infrastructure. These achievements underscore the importance of continued research and collaboration in the field of atmospheric science, as we strive to enhance our technological capabilities and resilience in an ever-changing space environment.

Achievements and Breakthroughs

The High-Frequency Active Auroral Research Program (HAARP) has achieved numerous breakthroughs and made significant contributions to scientific knowledge since its inception. This section outlines some of the most notable experiments, discoveries, and their implications for science and technology.

Artificial Aurora Generation

One of HAARP's most fascinating achievements is the generation of artificial auroras. By directing high-frequency (HF) radio waves into the ionosphere, HAARP can create localized plasma clouds that emit light, mimicking the natural auroral processes. These artificial auroras are not only visually striking but also scientifically invaluable.

The experiments to generate artificial auroras have provided critical insights into the mechanisms behind natural auroras, such as those observed near the polar regions. By studying these induced phenomena, scientists can better understand how charged particles from the sun interact with Earth's magnetic field and ionosphere. This knowledge is crucial for predicting and mitigating the effects of space weather events, such as geomagnetic storms, which can disrupt satellite communications and power grids.

The generation of artificial auroras at HAARP has also advanced our understanding of ionospheric physics. The controlled experiments allow researchers to observe the dynamics of plasma waves and their interactions with electromagnetic fields. These studies have revealed new information about ionospheric heating, plasma instabilities, and the behavior of energetic particles, contributing to the broader field of space physics.

Radio Wave Propagation Discoveries

Another significant area of HAARP's research involves the propagation of radio waves through the atmosphere. Understanding how radio waves travel and interact with the ionosphere is essential for improving communication and navigation technologies. HAARP's experiments have led to several key discoveries in this field.

One of the major findings is the behavior of HF radio waves under different ionospheric conditions. By conducting experiments during various phases of the solar cycle and under different geomagnetic conditions, HAARP has provided valuable data on how radio waves are reflected, refracted, and absorbed by the ionosphere. These insights have improved the accuracy of long-range communication systems, such as over-the-horizon radar, which rely on ionospheric reflection to detect objects beyond the line of sight.

HAARP's research has also led to advancements in understanding radio wave scattering and interference. The facility's experiments have shown how irregularities in the ionosphere, such as turbulence and plasma density fluctuations, can affect signal clarity and strength. These findings have practical applications in designing more robust communication systems that can maintain signal integrity under challenging conditions, such as during geomagnetic storms or in polar regions.

Plasma Physics Research

HAARP has made significant contributions to the field of plasma physics, particularly in understanding how electromagnetic waves interact with ionospheric plasma. Plasma, a state of matter consisting of charged particles, is ubiquitous in space and plays a critical role in many astrophysical processes.

One notable experiment conducted at HAARP involved the creation of artificial plasma clouds, or "plasma blobs," in the ionosphere. By manipulating the HF radio waves, researchers were able to study the formation, dynamics, and decay of these plasma structures. The findings from these experiments have enhanced our understanding of plasma transport and turbulence, which are fundamental to space weather phenomena.

HAARP has also investigated the interactions between electromagnetic waves and plasma instabilities. These studies have provided insights into phenomena such as the generation of plasma waves, their propagation, and their impact on ionospheric dynamics. Understanding these processes is crucial for developing predictive models of space weather and for designing systems that can withstand or exploit these interactions for practical applications.

In summary, HAARP's achievements and breakthroughs have significantly advanced our knowledge of atmospheric and ionospheric science. Through pioneering experiments on artificial auroras, radio wave propagation, and plasma physics, HAARP has contributed valuable insights that have practical applications in communication technologies, space weather prediction, and our overall understanding of the Earth's upper atmosphere. These achievements underscore the importance of continued research and innovation at HAARP and similar scientific facilities.

Future Research Directions

As the High-Frequency Active Auroral Research Program (HAARP) continues to evolve, its research agenda is set to explore new frontiers in atmospheric science, space weather, and beyond. This section outlines the emerging research areas, technological innovations, and long-term goals that will shape HAARP's future.

Emerging Research Areas

Looking ahead, HAARP is poised to delve into several promising research areas that build on its foundational work in ionospheric science. One emerging area of interest is advanced ionospheric modeling. By leveraging new data and improved computational techniques, researchers aim to develop more accurate models that can predict ionospheric behavior under various conditions. These models will be invaluable for forecasting space weather events and mitigating their impacts on communication and navigation systems.

Another promising research direction is the study of extreme atmospheric events. Understanding phenomena such as sudden stratospheric warmings, polar vortex disruptions, and other large-scale atmospheric dynamics can provide insights into their effects on the ionosphere and overall climate system. HAARP's unique capabilities make it an ideal platform for investigating these interactions, which could lead to better prediction and management of weather-related disruptions.

Additionally, HAARP is expected to expand its research on space weather impact mitigation. This involves studying how solar storms and geomagnetic disturbances affect modern technology, from satellite operations to power grid stability. Collaborations with other research institutions and international space agencies will be crucial in enhancing global space weather forecasting and developing strategies to protect critical infrastructure.

Technological Innovations

To support these ambitious research goals, HAARP plans to implement several technological innovations and upgrades. Enhancing the facility's equipment and capabilities will allow for more precise and comprehensive experiments, driving advancements in ionospheric research and beyond.

One significant upgrade is the development of new diagnostic instruments. These advanced tools will enable researchers to capture more detailed and accurate data on ionospheric conditions, including high-resolution measurements of plasma density, temperature, and electromagnetic wave interactions. With these improvements, HAARP can conduct more sophisticated experiments, yielding deeper insights into ionospheric processes.

Another innovation involves the integration of artificial intelligence (AI) and machine learning technologies. By analyzing vast amounts of data collected from HAARP's experiments, AI algorithms can identify patterns and trends that may not be apparent through traditional analysis methods. This approach will enhance our understanding of complex ionospheric phenomena and improve the predictive capabilities of space weather models.

Moreover, HAARP plans to upgrade its high-frequency transmitters to increase their power and versatility. These enhancements will allow for more diverse experimental setups, enabling researchers to explore a wider range of ionospheric conditions and phenomena. The improved transmitters will also facilitate collaborations with other research facilities, both within the United States and internationally, fostering a global approach to atmospheric and space weather research.

Long-Term Goals

HAARP's long-term goals are guided by its mission to advance scientific knowledge, improve communication technologies, and contribute to national and global security. The facility aims to be at the forefront of interdisciplinary research, addressing emerging scientific challenges and fostering innovation.

One of HAARP's strategic visions is to become a central hub for ionospheric and space weather research, collaborating with academic institutions, government agencies, and international partners. By sharing data and resources, HAARP can support a wide range of scientific endeavors, from basic research to applied technologies. This collaborative approach will drive scientific discovery and enhance the collective understanding of our planet's upper atmosphere.

Another long-term goal is to contribute to the resilience of global technological infrastructure. HAARP's research on space weather impact mitigation and communication technologies is critical for protecting vital systems from natural and human-made disruptions. By developing strategies to predict and manage space weather events,

HAARP aims to safeguard satellites, power grids, and other infrastructure that are essential for modern society.

Finally, HAARP is committed to promoting scientific literacy and public engagement. Through educational programs, public outreach, and transparent communication, HAARP seeks to demystify its research and foster a deeper appreciation for atmospheric science. By engaging with the public and sharing the excitement of scientific discovery, HAARP can inspire the next generation of scientists and build a more informed society.

In conclusion, HAARP's future research directions, technological innovations, and long-term goals highlight its dedication to advancing atmospheric science and addressing critical challenges. As HAARP continues to explore new frontiers, its contributions will have far-reaching implications for science, technology, and society. By embracing collaboration, innovation, and education, HAARP is poised to remain a leading institution in the study of the ionosphere and space weather, driving progress and understanding in these vital fields.

Chapter 12: HAARP in Popular Culture

HAARP in Movies and TV Shows

The High-Frequency Active Auroral Research Program (HAARP) has captured the public imagination, often depicted in movies and television shows as a mysterious and powerful entity. These portrayals, while often fictional and sensationalized, have significantly influenced how HAARP is perceived by the general public.

Movies Featuring HAARP

HAARP has made its way into the plotlines of several films, serving as a catalyst for suspense and intrigue. One notable example is the 2009 movie *The Objective*, where HAARP is depicted as a covert government project linked to supernatural events in Afghanistan. In the film, a special forces team encounters inexplicable phenomena, and the narrative suggests that HAARP's technology might be responsible. The film plays on themes of government secrecy and advanced technology, feeding into the conspiracy theories that already surround HAARP.

Another example is the science fiction thriller *The Signal* (2014), which involves a storyline where a mysterious signal disrupts electronic devices and communications. While not directly naming HAARP, the film alludes to similar high-frequency research facilities, drawing on the mystique and fear associated with such technology. These movies often amplify the aura of HAARP as an omnipotent entity capable of far-reaching and nefarious activities, perpetuating the myths rather than reflecting its real scientific objectives.

TV Shows and Documentaries

HAARP has also been featured in various television shows and documentaries, spanning genres from science fiction to investigative journalism. In the TV show *Conspiracy Theory with Jesse Ventura*, HAARP is examined through the lens of conspiracy theories, with episodes dedicated to exploring its supposed capabilities in mind control and weather manipulation. Ventura's show dramatizes these theories, presenting them as plausible scenarios and reinforcing public suspicion and intrigue.

Science fiction series like *The X-Files* have also touched upon HAARP-like themes. Although not explicitly naming HAARP, the show's episodes dealing with secret government projects and electromagnetic experiments echo the conspiracy theories associated with HAARP. These fictional portrayals contribute to the popular culture narrative that surrounds HAARP, blending elements of truth with speculative fiction to create compelling storylines.

Documentaries, on the other hand, approach HAARP from a more factual perspective, although some still carry an undercurrent of skepticism. Programs like the History Channel's *Decoding the Past* and National Geographic's *Naked Science* have featured segments on HAARP, discussing its scientific purposes while also acknowledging the conspiracy theories. These documentaries aim to provide a balanced view but often cannot completely dispel the myths, as the sensational aspects tend to captivate audiences.

Pop Culture Impact

The portrayal of HAARP in movies and TV shows has a profound impact on public perception. By positioning HAARP as a central element in stories of intrigue and danger, these media forms contribute to the mythos that surrounds the facility. The dramatic and often exaggerated depictions can overshadow the real scientific achievements and objectives of HAARP, embedding the idea of a shadowy, omnipotent organization in the public consciousness.

This blending of fact and fiction creates a challenge for those trying to convey HAARP's legitimate scientific work. While entertainment media is not beholden to the same standards of accuracy as scientific publications, its influence on public imagination is undeniable. The balance between entertainment and factual representation is delicate, and while the dramatization of HAARP may make for engaging stories, it also perpetuates misconceptions.

In summary, HAARP's portrayal in movies and television shows has significantly shaped its public image. These depictions often emphasize mystery and power, aligning with conspiracy theories rather than scientific reality. While these portrayals can be entertaining, they contribute to the persistent myths that surround HAARP, complicating efforts to accurately inform the public about the facility's true nature and purpose. As HAARP continues to be a topic of interest in popular culture, understanding and addressing these portrayals remains essential for fostering a more informed and realistic public perception.

HAARP in Books and Literature

The enigmatic nature of the High-Frequency Active Auroral Research Program (HAARP) has not only inspired screenwriters and filmmakers but has also found a prominent place in literature. Both fictional and non-fictional works have explored HAARP's potential, often blurring the lines between fact and fiction. This section delves into how HAARP is depicted in books and literature, examining the themes authors explore and the impact on public perception.

Fictional Works

HAARP has served as a compelling plot device in numerous novels, particularly within the science fiction and thriller genres. Authors often use HAARP's real-life capabilities as a springboard for imaginative and sometimes dystopian narratives. For instance, in the novel *Angels Don't Play This HAARP* by Nick Begich and Jeane Manning, the authors delve into speculative fiction based on real scientific principles, presenting HAARP as a tool for both beneficial and malevolent purposes. The book blends technical details with speculative scenarios, reinforcing the aura of mystery and power surrounding HAARP.

Similarly, in the techno-thriller *Polar Star* by Tom Clancy, HAARP is portrayed as a pivotal element in a geopolitical conflict. The novel's intricate plot uses HAARP's scientific capabilities to craft a narrative filled with suspense and high stakes, playing on themes of government secrecy and advanced technology. Such fictional portrayals contribute to the mythos of HAARP, captivating readers while often distorting the true nature of its scientific mission.

Non-Fictional Works

Non-fiction books about HAARP often aim to uncover the truth behind the facility's operations, sometimes leaning into investigative journalism and conspiracy theories. One of the most notable non-fiction books is also *Angels Don't Play This HAARP* by Nick Begich and Jeane Manning. While categorized as non-fiction, the book's speculative approach has fueled many conspiracy theories, suggesting that HAARP's capabilities extend far beyond publicly acknowledged scientific research.

Other non-fiction works take a more balanced view, presenting HAARP's history, objectives, and scientific achievements. Books like *The Pentagon's Brain* by Annie Jacobsen discuss HAARP in the context of broader military and scientific endeavors. Jacobsen's work provides a detailed account of various defense projects, including HAARP, offering insights into their development and objectives without veering into unfounded speculation. Such works contribute to a more nuanced understanding of HAARP, counteracting some of the more sensational claims.

Literary Influence

The presence of HAARP in both fictional and non-fictional literature significantly shapes public perception. Fictional works, with their dramatic and often dystopian narratives, captivate readers' imaginations and can perpetuate myths about HAARP's potential for causing global crises or being a tool for government control. These stories, while entertaining, often overshadow the factual scientific research conducted at HAARP.

Non-fiction works, on the other hand, play a dual role. While some books delve into speculative theories, contributing to the spread of misinformation, others strive to present a balanced and fact-based account of HAARP's activities. The impact of these books depends largely on their approach to the subject matter and the balance they strike between sensationalism and factual reporting.

Authors play a crucial role in shaping the narrative around HAARP, whether intentionally or unintentionally. By choosing to highlight certain aspects of HAARP's capabilities and operations, authors influence how readers perceive the facility. Fictional portrayals tend to amplify the mysterious and potentially ominous nature of HAARP, while well-researched non-fiction works aim to educate and inform the public about its legitimate scientific contributions.

In conclusion, HAARP's depiction in books and literature reflects the broader societal fascination with advanced scientific projects and government secrecy. Fictional works use HAARP to craft compelling narratives filled with intrigue and suspense, while non-fiction books offer varying perspectives, from speculative theories to balanced accounts of its scientific achievements. Together, these literary portrayals contribute to the complex and often misunderstood image of HAARP in popular culture, highlighting the need for ongoing efforts to separate fact from fiction and promote a more accurate understanding of its true purpose and capabilities.

HAARP in Online Communities and Social Media

In the digital age, online communities and social media platforms have become fertile ground for discussions about the High-Frequency Active Auroral Research Program (HAARP). These platforms amplify voices, spread information rapidly, and often blur the lines between fact and fiction. This section explores HAARP's presence in online forums, discussion boards, and social media, examining the impact of these digital spaces on public perception.

Online Forums and Discussion Boards

Online forums and discussion boards are hotspots for HAARP-related discussions, where users engage in lively debates, share theories, and speculate about the facility's capabilities. Platforms like Reddit, AboveTopSecret, and Godlike Productions have dedicated threads discussing HAARP, often from a conspiratorial perspective. These forums allow for anonymous participation, which can encourage the sharing of controversial and unverified information without the accountability of identified authorship.

In these digital spaces, discussions often revolve around the potential for HAARP to influence weather, control minds, or cause natural disasters. Users share articles, videos, and personal anecdotes, creating a collective narrative that reinforces their beliefs. The anonymity and community dynamics of these forums can lead to echo chambers, where like-minded individuals validate each other's suspicions, further entrenching conspiracy theories.

The role of moderation in these forums varies. Some platforms actively moderate discussions to prevent the spread of misinformation, while others take a more laissez-faire approach, allowing unverified claims to proliferate. The impact of these discussions on public perception is significant, as they reach a wide audience and can shape the views of those who might not seek out alternative perspectives.

Social Media Platforms

Social media platforms like Facebook, Twitter, and YouTube play a crucial role in spreading information about HAARP, both accurate and inaccurate. These platforms allow users to share content with vast audiences, often with the potential for rapid viral spread. Videos, posts, and memes about HAARP can quickly gain traction, reaching millions of people and shaping public opinion.

On YouTube, for example, numerous videos claim to expose the "truth" about HAARP, often presenting speculative theories as fact. These videos use dramatic visuals, ominous music, and persuasive narration to captivate viewers and suggest hidden dangers. Channels dedicated to conspiracy theories frequently feature HAARP, reinforcing the narrative of a secretive and powerful entity.

Twitter and Facebook also host a myriad of discussions about HAARP. Hashtags related to HAARP can trend, bringing the topic to a broader audience and encouraging further debate. Posts range from legitimate scientific explanations to sensationalist and conspiratorial claims. The algorithms of these platforms, which prioritize engaging and shareable content, can inadvertently amplify the most dramatic and attention-grabbing posts, regardless of their factual accuracy.

Digital Influence

The influence of online and social media discussions on public attitudes towards HAARP is profound. These platforms democratize information dissemination, allowing anyone to contribute to the narrative, but they also pose challenges in combating misinformation. The sheer volume of content and the ease with which false information can spread make it difficult to ensure the accuracy of public discourse.

Combating misinformation in digital spaces requires a multifaceted approach. Enhancing digital literacy is crucial, teaching users to critically evaluate sources, verify facts, and recognize biases. Platforms themselves have a role to play in promoting accurate information and curbing the spread of falsehoods. Fact-checking organizations and partnerships with scientific institutions can help provide reliable information and counteract misleading claims.

Scientific and educational communities must also engage with these digital platforms to provide accurate and accessible information about HAARP. By participating in discussions, sharing verified data, and correcting misconceptions, experts can help steer the narrative towards a more factual understanding of HAARP's activities.

In conclusion, HAARP's presence in online communities and social media significantly impacts public perception. While these platforms enable widespread discussion and democratize information, they also facilitate the spread of misinformation. Addressing this challenge requires a concerted effort to promote digital literacy, engage with credible sources, and ensure accurate information prevails in the digital discourse surrounding HAARP. By doing so, we can foster a more informed and rational public understanding of HAARP and its scientific contributions.

Chapter 13: Conclusion: Separating Fact from Ficti

Summary of Key Points

As we delve into the heart of this book, it's essential to revisit the myriad insights and revelations unearthed in our journey. This section will illuminate the salient points discussed, tying together critical analyses, scientific evidence, and rational arguments presented in each chapter.

Revisiting HAARP's Purpose and Research

The High-Frequency Active Auroral Research Program (HAARP) was established with a noble vision to further our understanding of the Earth's ionosphere, an endeavor that has significantly contributed to atmospheric science, ionospheric research, and space weather monitoring. Through rigorous scientific inquiry, HAARP has provided invaluable data on the ionosphere's influence on radio wave propagation, which is critical for communication technologies.

Key experiments conducted under HAARP have led to breakthroughs in comprehending the ionosphere's complex dynamics. From artificially generating small-scale ionospheric disturbances to studying natural phenomena like auroras, these efforts have culminated in a more profound grasp of space weather's impact on satellite operations, GPS systems, and global communication networks.

Debunking Conspiracy Theories

The allure of conspiracy theories often stems from a blend of fear, curiosity, and misinformation. Throughout this book, we have systematically dissected and debunked the most prominent myths surrounding HAARP, including allegations of weather control, mind control, and geological manipulation.

By presenting evidence-based arguments and scientific rebuttals, we have dispelled the notion that HAARP possesses any capability to manipulate weather patterns or influence human behavior. Instead, we highlighted how these conspiracy theories have arisen from a misunderstanding of HAARP's legitimate research activities and a tendency to ascribe nefarious intentions to advanced scientific endeavors.

Media and Public Perception

HAARP's portrayal in media, popular culture, and online communities has been a double-edged sword. While it has drawn attention to the importance of ionospheric research, it has also been the subject of sensationalist portrayals that have skewed public perception and fueled the spread of misinformation.

This book has scrutinized the role of media in shaping HAARP's image, emphasizing how sensationalist reporting can often overshadow nuanced scientific discourse. By understanding the media's influence, we can better appreciate the need for accurate and responsible communication of scientific research to the public.

In summary, this book has sought to clarify HAARP's true mission and achievements, debunk unfounded conspiracy theories, and critically examine the media's role in shaping public opinion. Through this comprehensive exploration, we aim to foster a more informed and rational understanding of HAARP's contributions to science and society.

Importance of Critical Thinking and Scientific Literacy

The bedrock of any meaningful inquiry into complex topics such as HAARP lies in the ability to think critically and possess a robust level of scientific literacy. These attributes are indispensable tools that empower individuals to evaluate theories and claims with a discerning eye, separating fact from fiction.

Understanding Scientific Principles

A thorough grounding in scientific principles forms the backbone of evaluating extraordinary claims. Science operates on the bedrock of evidence, experimentation, and peer review. Hence, understanding the fundamentals of scientific inquiry is crucial for anyone seeking to make sense of controversial topics.

Readers are encouraged to delve into credible scientific sources and peer-reviewed research. By doing so, they can equip themselves with accurate information and a deeper appreciation of the rigorous processes that underpin scientific discoveries. This approach fosters a more informed perspective, enabling individuals to distinguish well-substantiated facts from speculative assertions.

Developing Critical Thinking Skills

Critical thinking is the art of evaluating information with a healthy dose of skepticism and open-mindedness. It involves questioning the credibility of sources, scrutinizing evidence, and recognizing biases that may cloud judgment.

A key element of critical thinking is understanding that not all correlations indicate causation. Just because two events occur simultaneously does not mean one caused the other. Recognizing logical fallacies—such as false dilemmas, slippery slopes, and ad hominem attacks—can help individuals navigate through flawed arguments and arrive at more rational conclusions.

Practical tips for honing critical thinking skills include:

- **Questioning Sources:** Always consider the reliability and credibility of the information source.
- **Verifying Evidence:** Cross-check facts and data from multiple reputable sources.
- **Recognizing Biases:** Be aware of personal and external biases that might influence interpretations.
- **Understanding Context:** Place information within the broader context to see the bigger picture.

Promoting Scientific Literacy

A scientifically literate society is better equipped to make informed decisions and engage in rational discourse. Comprehensive science education and public outreach initiatives play a pivotal role in enhancing scientific literacy.

Educators, scientists, and media professionals have a crucial responsibility in disseminating accurate information and countering misinformation. By fostering an environment that values scientific inquiry and critical thinking, these stakeholders can contribute to a more informed public that can engage with scientific topics constructively.

Furthermore, a scientifically literate populace benefits society at large. It encourages informed decision-making in personal and public spheres, promotes rational debate, and helps to build a culture that values evidence-based reasoning over conjecture and speculation.

In essence, this section underscores the vital role of critical thinking and scientific literacy in navigating the complexities of topics like

HAARP. By cultivating these skills, individuals can better evaluate theories and claims, contributing to a more informed and rational discourse.

Call for Rational Discourse and Continued Scientific Inquiry

As we reach the conclusion of this exploration into the world of HAARP, it is imperative to recognize the importance of engaging in rational discourse and supporting the continual advancement of scientific inquiry. This final section serves as a call to action, urging readers to approach discussions with an open mind and a commitment to evidence-based reasoning, while also emphasizing the significance of sustained scientific endeavors.

Encouraging Rational Discourse

In the realm of scientific inquiry, the exchange of ideas and respectful dialogue are paramount. Readers are encouraged to participate in discussions about HAARP and similar topics with an open mind, actively seeking out credible sources and prioritizing evidence-based reasoning. By fostering an environment where ideas can be shared and debated constructively, we can work towards a more informed and understanding society.

The value of respectful dialogue cannot be overstated. It is through the exchange of diverse perspectives that we can challenge assumptions, refine our understanding, and arrive at more nuanced conclusions. Rational discourse, grounded in a commitment to truth and objectivity, is essential for the progress of knowledge.

Supporting Scientific Inquiry

Continued investment in scientific research and innovation is critical for the advancement of knowledge and technology. Projects like HAARP exemplify the importance of sustained scientific inquiry, providing valuable insights into atmospheric and ionospheric phenomena that have far-reaching implications for communication technologies and space weather monitoring.

Public support plays a vital role in funding and sustaining scientific endeavors. Advocacy for scientific research and innovation ensures that groundbreaking projects receive the resources needed to thrive. By supporting scientific inquiry, we contribute to the collective pursuit of knowledge and the betterment of society.

Encouraging a culture of curiosity is equally important. Readers are urged to stay curious, ask questions, and seek out new knowledge. This proactive engagement with the world around us fosters a deeper understanding and appreciation of the complexities of our universe. By remaining inquisitive, we can contribute to the ongoing dialogue that drives scientific progress.

Reflecting on the Journey

As we conclude this book, it is worth reflecting on the journey we have undertaken. Consider how your perspectives may have evolved and how the insights gained from this exploration have shaped your understanding of HAARP and its scientific contributions.

Scientific discovery is an ongoing process, characterized by a continuous quest for knowledge and a vigilance against misinformation. It is essential to remain open to new information, question assumptions, and adapt our understanding as new evidence emerges.

In closing, let us embrace the power of science and critical thinking to illuminate the truth and drive progress. By fostering a culture of rational discourse and supporting scientific inquiry, we can collectively contribute to a brighter, more informed future. Together, we can navigate the complexities of our world with curiosity, rigor, and a steadfast commitment to truth.

Chapter 14: Appendices and References

Additional Data, Charts, and Technical Details Related to HAARP

In the pursuit of scientific clarity, this section delves into the wealth of additional data, charts, and technical details that underpin the High-Frequency Active Auroral Research Program (HAARP). These appendices serve as a comprehensive repository of information, offering readers an in-depth look at the empirical foundation of HAARP's research activities.

Introduction to Appendices

To fully appreciate the scope and impact of HAARP's work, it is essential to examine the supplementary data and technical details that support the main content of this book. This section outlines the specific types of data and charts included, providing a roadmap for readers to navigate the intricate details of HAARP's research.

Detailed Data Sets

The appendices feature a collection of data sets derived from various HAARP experiments. These data sets encompass a wide range of measurements and observations, offering a granular view of the ionosphere's behavior and its interactions with radio waves.

Methodology: The data presented here was meticulously gathered through a series of controlled experiments. Researchers employed sophisticated instrumentation to capture real-time measurements of ionospheric parameters, such as electron density, temperature, and plasma turbulence. The methodology section provides a detailed account of the experimental procedures, ensuring transparency and reproducibility.

Results and Significance: The interpretation of these data sets reveals critical insights into the ionosphere's dynamics. By analyzing the results, scientists have been able to identify patterns and anomalies that enhance our understanding of space weather and its impact on communication technologies. The significance of these findings extends to practical applications, including the improvement of satellite communication systems and the mitigation of space weather-related disruptions.

Technical Specifications

To comprehend the full scope of HAARP's capabilities, it is essential to explore the technical specifications of the equipment and technology used in its experiments. This section provides detailed descriptions and diagrams of the technical setup, offering readers a behind-the-scenes look at the infrastructure that powers HAARP's research.

Equipment and Technology: The appendices include comprehensive information on the various instruments and devices utilized in HAARP's experiments. From high-frequency transmitters to sophisticated radar systems, each piece of equipment is described in detail, with accompanying charts and diagrams to illustrate their functions.

Technical Setup: Diagrams and flowcharts offer a visual representation of the experimental setup, showcasing the layout and configuration of the equipment. These visual aids help readers grasp the complexity of HAARP's research environment and the meticulous planning involved in conducting experiments.

Contributions to Understanding: The technical specifications section underscores how these details contribute to the broader understanding of HAARP's research. By examining the equipment and methodology, readers can appreciate the precision and rigor that characterize HAARP's scientific endeavors.

Comprehensive List of References

In any scholarly work, the foundation of credibility and depth is supported by a thorough and meticulously curated list of references. This section presents a comprehensive compilation of the sources that have informed and enriched our exploration of HAARP. By offering a well-organized and annotated bibliography, we aim to provide readers with a valuable resource for further investigation and validation of the information presented in this book.

Introduction to References

The importance of a comprehensive list of references cannot be overstated. References serve as the bedrock of scholarly work, allowing readers to trace the origins of ideas, verify facts, and delve deeper into the topics covered. By providing a detailed and organized list of references, we uphold the standards of academic rigor and transparency, ensuring that our work is both credible and informative.

Categorization of References

To facilitate ease of navigation, the references are categorized into distinct sections. This categorization allows readers to quickly locate sources of interest and explore specific aspects of HAARP and its related scientific fields. The main categories include:

- **Scientific Papers**: Peer-reviewed journal articles that provide empirical data, experimental results, and theoretical analyses related to HAARP.
- **Technical Reports**: Comprehensive reports detailing the technical aspects and methodologies of HAARP experiments and research projects.
- **Books**: Authoritative texts that offer in-depth discussions on atmospheric science, ionospheric research, and the broader context of HAARP's contributions.
- **Articles**: Informative pieces from reputable publications that cover various facets of HAARP and its public perception.

Annotated Bibliography

An annotated bibliography enhances the utility of the references by offering brief summaries and explanations of each source's content and relevance. This section includes annotations for key references, providing readers with a snapshot of the information contained within each source and its significance to the overall narrative of the book.

- **Scientific Papers**:
    - *Author(s)*: Title of the paper. *Journal Name*, Volume(Issue), Page numbers. Brief summary of the study's focus, methodology, and key findings, highlighting its relevance to HAARP.
- **Technical Reports**:
    - *Author(s)*: Title of the report. *Institution/Organization*. Summary of the technical details, experimental setup, and outcomes discussed in the report, emphasizing its contribution to understanding HAARP's research.
- **Books**:

- *Author(s)*: Title of the book. Publisher, Year. Overview of the book's content, the author's background, and its relevance to the topics covered in this book.
- **Articles**:
  - *Author(s)*: Title of the article. *Publication Name*, Date. Summary of the article's main points, the context in which it was written, and its significance to the discussion of HAARP.

By categorizing and annotating the references, we provide readers with a clear and accessible pathway to further explore the wealth of knowledge that has shaped our understanding of HAARP. This comprehensive list serves as a testament to the extensive research and scholarly diligence that underpins the content of this book.

Further Reading and Resources

In the pursuit of deeper understanding and continued exploration, this section provides a curated list of further reading and resources. By engaging with these materials, readers can expand their knowledge of HAARP and related scientific fields. This compilation includes recommended books, articles, online resources, and multimedia, offering diverse avenues for continued learning.

Introduction to Further Reading

The journey of scientific discovery does not end with the conclusion of this book. Readers are encouraged to seek out additional resources to deepen their understanding of the topics covered and to remain curious about the world around them. This section serves as a guide to further exploration, presenting a selection of reputable sources that can enhance the reader's knowledge and engagement with HAARP and its broader scientific context.

Recommended Books and Articles

A wealth of literature exists on the subjects of atmospheric science, ionospheric research, and space weather. The following books and articles are particularly valuable for their insights and contributions to these fields:

- **Books**:
    - *Author(s)*: Title of the book. Publisher, Year. This book provides an in-depth exploration of [topic], offering a comprehensive overview of [related subjects].
    - *Author(s)*: Title of the book. Publisher, Year. A seminal work in the field of [subject], this book delves into the intricacies of [specific topic], making it a must-read for those interested in [related areas].
- **Articles**:
    - *Author(s)*: Title of the article. *Publication Name*, Date. This article presents a detailed analysis of [subject], highlighting recent advancements and key findings.
    - *Author(s)*: Title of the article. *Publication Name*, Date. An insightful piece that explores [related topic], providing valuable context and understanding of [subject].

Online Resources and Multimedia

In the digital age, a plethora of online resources and multimedia materials are available to complement traditional reading. These resources offer dynamic and interactive ways to engage with scientific content:

- **Websites**:

- ◦ **Website Name**: URL. A reputable website that offers a wealth of information on [subject], including articles, research papers, and interactive tools.
- ◦ **Website Name**: URL. This online database provides access to a vast collection of scientific publications, allowing readers to explore cutting-edge research in [field].
- **Multimedia**:
  - ◦ **Video Series**: Title of the series. Available on [Platform]. This video series offers visual explanations and demonstrations of [subject], making complex concepts accessible and engaging.
  - ◦ **Podcast**: Title of the podcast. Available on [Platform]. A thought-provoking podcast that features discussions with experts in [field], providing diverse perspectives and insights.

Conclusion and Call to Action

As we reach the final pages of this book, it is important to recognize that the pursuit of knowledge is an ongoing journey. Readers are encouraged to remain curious, ask questions, and seek out new information. By engaging with the recommended resources and participating in discussions, readers can contribute to the collective understanding of HAARP and its scientific endeavors.

The power of science and critical thinking lies in their ability to illuminate the truth and drive progress. By fostering a culture of rational discourse and supporting continued scientific inquiry, we can navigate the complexities of our world with curiosity, rigor, and a steadfast commitment to evidence-based reasoning.

In closing, let us celebrate the spirit of inquiry and the pursuit of knowledge. Through continued exploration and engagement, we can collectively contribute to a more informed and understanding society, where the power of science and critical thinking guides us towards a brighter future.